Sport, Ethics, and Neurophilosophy

The growth of neuroscience and the spread of general interest in the brain have prompted concern for ethical issues posed by neuroscientists. Despite the growing interest in the brain, neuroscience, and the profound issues that neuroscience raises, up to this point, relatively little attention has been given to, broadly speaking, neurophilosophical reflection on the brain in the context of sport. This book seeks to address this gap.

Sport abounds with issues ripe for neurophilosophical treatment. Human movement, intentionality, cognition, cooperation, and vulnerability to injury directly and indirectly implicate the brain, and feature prominently in sport. This innovative volume comprises chapters by a team of international scholars who have written on a wide variety of topics at the intersection of sport, ethics, and neurophilosophy. Not only are the issues presented here of pressing philosophical and practical concerns, but they also represent a new mode of fluid interaction between science and philosophy for the future of sports scholarship.

This book was originally published as a special issue of the journal *Sport, Ethics and Philosophy*.

Jeffrey P. Fry is an Associate Professor in the Department of Philosophy and Religious Studies at Ball State University, Muncie, USA. His recent research lies at the intersection of sport and neurophilosophy. He formerly served on the Executive Council of the International Association for the Philosophy of Sport.

Mike McNamee is a Professor of Applied Ethics at Swansea University, UK. He is the author of *Bioethics, Genetics and Sport* (2018, with Silvia Camporesi), and co-edits the *Ethics and Sports* book series (Routledge). He is a member of WADAs Ethics Committee, a former President of the International Association for the Philosophy of Sport, and the Founding Editor of *Sport, Ethics and Philosophy*.

Ethics and Sport
Series editors: *Mike McNamee, University of Wales Swansea*
Jim Parry, University of Leeds

The *Ethics and Sport* series aims to encourage critical reflection on the practice of sport, and to stimulate professional evaluation and development. Each volume explores new work relating to philosophical ethics and the social and cultural study of ethical issues. Each is different in scope, appeal, focus and treatment but a balance is sought between local and international focus, perennial and contemporary issues, level of audience, teaching and research application, and variety of practical concerns.

Recent titles include:

Doping in Elite Sports
Voices of French Sportspeople and Their Doctors, 1950–2010
Christophe Brissonneau and Jeffrey Montez de Oca

Bioethics, Genetics and Sport
Silvia Camporesi and Mike McNamee

Body Ecology and Emersive Leisure
Edited by Bernard Andrieu, Jim Parry, Alessandro Porrovecchio and Olivier Sirost

Sport, Ethics and Philosophy
Edited by Mike McNamee

Philosophy and Nature Sports
Kevin Krein

Emotion in Sports
Philosophical Perspectives
Yunus Tuncel

Sport, Ethics, and Neurophilosophy
Jeffrey P. Fry and Mike McNamee

Doping in Cycling
Interdisciplinary Perspectives
Edited by Bertrand Fincoeur, John Gleaves and Fabien Ohl

For a complete series list please visit: https://www.routledge.com/Ethics-and-Sport/book-series/EANDS

Sport, Ethics, and Neurophilosophy

Edited by
Jeffrey P. Fry and Mike McNamee

LONDON AND NEW YORK

First published 2019
by Routledge
2 Park Square, Milton Park, Abingdon, Oxon, OX14 4RN, UK

and by Routledge
52 Vanderbilt Avenue, New York, NY 10017, USA

First issued in paperback 2020

Routledge is an imprint of the Taylor & Francis Group, an informa business

Chapters 1–5, 7–11© 2019 Taylor & Francis
Chapter 6 © 2017 Hub Zwart. Originally published as Open Access.

With the exception of Chapter 6, no part of this book may be reprinted or reproduced or utilised in any form or by any electronic, mechanical, or other means, now known or hereafter invented, including photocopying and recording, or in any information storage or retrieval system, without permission in writing from the publishers. For details on the rights for Chapter 6, please see the chapter's Open Access footnote.

Trademark notice: Product or corporate names may be trademarks or registered trademarks, and are used only for identification and explanation without intent to infringe.

British Library Cataloguing-in-Publication Data
A catalogue record for this book is available from the British Library

ISBN 13: 978-0-367-58289-0 (pbk)
ISBN 13: 978-1-138-39395-0 (hbk)

Typeset in Myriad Pro
by codeMantra

Publisher's Note
The publisher accepts responsibility for any inconsistencies that may have arisen during the conversion of this book from journal articles to book chapters, namely the possible inclusion of journal terminology.

Disclaimer
Every effort has been made to contact copyright holders for their permission to reprint material in this book. The publishers would be grateful to hear from any copyright holder who is not here acknowledged and will undertake to rectify any errors or omissions in future editions of this book.

Contents

Citation Information	vii
Notes on Contributors	ix

1 Sport, Ethics, and Neurophilosophy 1
  Jeffrey P. Fry and Mike McNamee

2 Ethics, Brain Injuries, and Sports: Prohibition, Reform, and Prudence 6
  Francisco Javier Lopez Frias and Mike McNamee

3 Governing sporting brains: concussion, neuroscience, and the biopolitical regulation of sport 23
  Jennifer Hardes

4 Two Kinds of Brain Injury in Sport 36
  Jeffrey P. Fry

5 On the Compatibility of Brain Enhancement and the Internal Values of Sport 49
  Alberto Carrio Sampedro and José Luis Pérez Triviño

6 Skiing and its Discontents: Assessing the *Turist* Experience from a Psychoanalytical, a Neuroscientific and a Sport Philosophical Perspective 65
  Hub Zwart

7 Intentional and Skillful Neurons 81
  Jens Erling Birch

8 Team Spirit, Team Chemistry, and Neuroethics 99
  Andrew Fiala

9 High-level Enactive and Embodied Cognition in Expert Sport Performance 112
  Kevin Krein and Jesús Ilundáin-Agurruza

CONTENTS

10 Neuropsychology Behind the Plate 127
    Jordan Edmund DeLong

11 Appendix: An Interview with Leonardo Fogassi 138
    Jens Erling Birch

    Index 153

Citation Information

The chapters in this book were originally published in the journal *Sport, Ethics and Philosophy*, volume 11, issue 3 (August 2017). When citing this material, please use the original page numbering for each article, as follows:

Chapter 1
Sport, Ethics, and Neurophilosophy
Jeffrey P. Fry and Mike McNamee
Sport, Ethics and Philosophy, volume 11, issue 3 (August 2017) pp. 259–263

Chapter 2
Ethics, Brain Injuries, and Sports: Prohibition, Reform, and Prudence
Francisco Javier Lopez Frias and Mike McNamee
Sport, Ethics and Philosophy, volume 11, issue 3 (August 2017) pp. 264–280

Chapter 3
Governing sporting brains: concussion, neuroscience, and the biopolitical regulation of sport
Jennifer Hardes
Sport, Ethics and Philosophy, volume 11, issue 3 (August 2017) pp. 281–293

Chapter 4
Two Kinds of Brain Injury in Sport
Jeffrey P. Fry
Sport, Ethics and Philosophy, volume 11, issue 3 (August 2017) pp. 294–306

Chapter 5
On the Compatibility of Brain Enhancement and the Internal Values of Sport
Alberto Carrio Sampedro and José Luis Pérez Triviño
Sport, Ethics and Philosophy, volume 11, issue 3 (August 2017) pp. 307–322

Chapter 6
Skiing and its Discontents: Assessing the Turist *Experience from a Psychoanalytical, a Neuroscientific and a Sport Philosophical Perspective*
Hub Zwart
Sport, Ethics and Philosophy, volume 11, issue 3 (August 2017) pp. 323–338

Chapter 7
Intentional and Skillful Neurons
Jens Erling Birch
Sport, Ethics and Philosophy, volume 11, issue 3 (August 2017) pp. 339–356

Chapter 8
Team Spirit, Team Chemistry, and Neuroethics
Andrew Fiala
Sport, Ethics and Philosophy, volume 11, issue 3 (August 2017) pp. 357–369

Chapter 9
High-level Enactive and Embodied Cognition in Expert Sport Performance
Kevin Krein and Jesús Ilundáin-Agurruza
Sport, Ethics and Philosophy, volume 11, issue 3 (August 2017) pp. 370–384

Chapter 10
Neuropsychology Behind the Plate
Jordan Edmund DeLong
Sport, Ethics and Philosophy, volume 11, issue 3 (August 2017) pp. 385–395

Chapter 11
Appendix: An Interview with Leonardo Fogassi
Jens Erling Birch
Sport, Ethics and Philosophy, volume 11, issue 3 (August 2017) pp. 396–410

For any permission-related enquiries please visit:
http://www.tandfonline.com/page/help/permissions

Notes on Contributors

Jens Erling Birch is an Associate Professor of Physical Education in the Faculty of Teacher Education and International Studies at Hogskolen i Oslo og Akershus, Norway.

Alberto Carrio Sampedro is a Lecturer in Philosophy of Law at Pompeu Fabra University, Barcelona, Spain. He has published several articles on the theory of law, constitutional law, the rule of law, sports rules, equality in sports, and the impact of cognitive improvement on the concept of justice.

Jordan Edmund DeLong is the Director of Data Science and Product Innovation for Research Narrative, a market research firm based out of Los Angeles, USA. He utilizes psychological theory, machine learning, and research design in his work.

Andrew Fiala is the Chair and a Professor of Philosophy, and the Director of the Ethics Center, at California State University, Fresno, USA. His scholarly writing focuses on war and peace, politics, religion, and ethics. He is also a columnist for *The Fresno Bee*.

Jeffrey P. Fry is an Associate Professor in the Department of Philosophy and Religious Studies at Ball State University, Muncie, USA. His recent research lies at the intersection of sport and neurophilosophy. He formerly served on the Executive Council of the International Association for the Philosophy of Sport.

Jennifer Hardes is an Academic Research Lead in the Research and Enterprise Development Centre; a Senior Lecturer in Social Philosophy, Law & Bioethics; and the Director of Postgraduate Research Training and Development at Canterbury Christ Church University, UK.

Jesús Ilundáin-Agurruza is a Professor and the Chair of Philosophy at Linfield College, McMinnville, USA. His academic interests range from value theory (ethics and aesthetics) to the philosophy of mind; from East Asian Philosophy (especially Japanese) to philosophy and literature, 20th-century philosophy, and metaphysics – all of which he enjoys relating to sport and related practices such as martial arts.

Kevin Krein is a Professor of Philosophy in the Department of Humanities at the University of Alaska Southeast, Juneau, USA. His primary philosophical work is in the areas of philosophy of nature and the environment and philosophy of mind.

Francisco Javier Lopez Frias is an Assistant Professor of Kinesiology and a Research Associate in the Rock Ethics Institute at Pennsylvania State University, State College, USA. He researches sports ethics and human enhancement but is also interested in political philosophy, normative ethics, and applied ethics.

NOTES ON CONTRIBUTORS

Mike McNamee is a Professor of Applied Ethics at Swansea University, UK. He is the author of *Bioethics, Genetics and Sport* (2018, with Silvia Camporesi), and co-edits the *Ethics and Sports* book series (Routledge). He is a member of WADAs Ethics Committee, a former President of the International Association for the Philosophy of Sport, and the Founding Editor of *Sport, Ethics and Philosophy*.

José Luis Pérez Triviño is a Senior Lecturer in Philosophy of Law and an Associate Professor at Pompeu Fabra University, Barcelona, Spain. He is the President of the Spanish Association of Philosophy of Sport and the Director of *Fair Play Journal*.

Hub Zwart is a Professor of Philosophy in the Faculty of Science at Radboud University, Nijmegen, the Netherlands. In 2004, he established the Centre for Society and Genomics and, in 2005, he established the Institute for Science, Innovation and Society.

Sport, Ethics, and Neurophilosophy

Jeffrey P. Fry and Mike McNamee

ABSTRACT
The influence of neuroscience looms large today. In this introductory essay, we provide some context for the volume by acknowledging the expansion of applied neuroscience to everyday life and the proliferation of neuroscientific disciplines. We also observe that some individuals have sounded cautionary notes in light of perceived overreach of some claims for neuroscience. Then we briefly summarize the articles that comprise this volume. This diverse collection of papers represents the beginning of a conversation focused on the intersection of sport, ethics and neurophilosophy. As such, the essays also represent a new mode of interaction between philosophy and science with sport in the center of the discussion.

Introduction

While outsiders might consider neuroscience a newcomer to the academy, Farah (2013, 762) notes that the seeds of the subject were planted as long ago as the 4th millennium BCE, when Sumerians noted the effects of the poppy plant on mood. Farah (2013, 762–3) distinguishes three kinds of advances in the field. The first two occurred between 4000 BCE and 2000 CE. The initial advance involved progress in understanding the nervous system. The second advance was in applied neuroscience, especially in the field of medicine. Now in the twenty-first century we are in the midst of a third neuroscientific advance, beyond the boundaries of medicine to everyday contexts of 'home, office, school, courtroom, marketplace, and battlefield' (Farah 2013, 763). The expansion of the applied influence of neuroscience has been accompanied by a proliferation of neuroscientific disciplines, such as 'neurolaw, neurophilosophy, neuromarketing … neurofinance … neuroaesthetics, neurohistory, neuroliterature, neuromusicology, neuropolitics, and neurotheology' (Satel and Lilienfeld 2013, ix). The Decade of the Brain, proclaimed by US President George H.W. Bush,[1] arrived in the 1990s, followed by the Brain Initiative and the European Union's Human Brain Project. As with many scientific advances, the lexicon slides into common usage, often via journalistic efforts. Thus, words like 'neuroplasticity', 'mirror neurons', and 'frontal lobes' appear in greater currency (with or without corresponding clarity and comprehension).

The academic community has not accepted each of the corresponding advances of neuroscience uncritically. An indictment of perceived excesses can be seen in the charges of 'neuromania', 'neurohubris', and 'neurobollocks' (Satel and Lilienfeld 2013, xiv). The backlash has brought criticisms of unwarranted neuoreductionism (see e.g. Glannon 2011). Yet despite this scorn for some of the aspirational claims of neuroscientists and other brain enthusiasts, interest in the brain waxes.

The growth of neuroscience and the spread of general interest in the brain have also sparked concerns for ethical issues raised by neuroscience. Racine (2010, 29) traces the term 'neuroethics' to Harvard physician Annelise A. Pontius in the 1970s. Pontius employed the designation while expressing concern for measures utilized to speed up the process through which infants mastered movement and the ability to walk. *New York Times* journalist William Safire is credited with giving usage of the term 'neuroethics' a more recent impetus. In a 2002 meeting in California entitled 'Neuroethics: Mapping the Field Conference', Safire described neuroethics 'as a distinct portion of bioethics, which is the consideration of good and bad consequences in medical practice and biological research' (quoted in Racine 2010, 30). One might question the reduction of ethical concerns to a consequentialist agenda, but one should also note the direction of travel from speculation to serious ethical concern. Also in 2002, the philosopher Adina Roskies (2002; see also Racine 2010, 31) helped delineate the field of neuroethics by distinguishing between two trajectories of the discipline: 'the ethics of neuroscience' and the 'neuroscience of ethics'. Roskies argues that this twofold agenda establishes neuroethics as a distinctive discipline that deserves to be singled out within the broader field of bioethics. The philosopher Neil Levi (2007) distinguishes between the ethics of neuroscience and the neuroscience of ethics as follows:

> The *ethics of neuroscience* refers to the branch of neuroethics that seeks to develop an ethical framework for regulating the conduct of neuroscientific enquiry and the application of neuroscientific knowledge to human beings; *the neuroscience of ethics* refers to the impact of neuroscientific knowledge upon our understanding of ethics itself. (Levi 2007, 1)

Despite the proliferation of interest in the brain, neuroscience, and the profound ethical issues that neuroscience raises, to date relatively little attention has been paid to broadly speaking neurophilosophical reflection on the brain in the context of sport.[2] As this issue goes to print, we are aware of no special issue or book that undertakes this task. This collection of essays seeks to address this lacuna. It is somewhat surprising that it has taken this long to address the brain in sport in this manner, since sport abounds with issues ripe for neurophilosophical treatment. Human movement, intentionality, cognition, cooperation, and vulnerability all implicate the brain, and all are prominently featured in the world of sport. In addition, ethical issues involving the brain are pervasive in sport, including, but not limited to, the pressing matter of concussions. Therefore, sensing a gap in the philosophy of sport literature, the editors of this volume have brought together a team of international scholars, who have written on a wide variety of topics at the intersection of sport, ethics, and neurophilosophy.

The issue begins with two essays that deal with the pressing issue of concussions. The first essay of this volume, by Mike McNamee and Francisco Javier Lopez Frias, is entitled 'Ethics, Brain Injuries and Sports: Prohibition, Reform, and Prudence'. The discussions around brain injuries in sport in general, and sports related concussion in particular, have raged in sports medicine for the last several years. The latter has been the object of several consensus statements by leading neuroscientists and sports medicine clinicians and scientists, notably from the Concussion in Sport Group (CISC). Lopez Frias and McNamee focus their attention on the work of three leading scholars who have either called for the reform of boxing (Dixon

2001), or the elimination/prohibition of American football (Corlett 2014; Pam Sailors 2015, respectively) on the basis of harm resulting from head injuries in sports. Lopez Frias and McNamee argue that the paternalistic stance in each case is not merited by the current evidential basis, and the weakness of arguments concerning the protection of the future autonomy of athletes that will be compromised by their athletic participation.

The second paper on sport related concussions is Jennifer Hardes' 'Governing Sporting Brains: Concussion, Neuroscience, and the Biopolitical Regulation of Sport'. Hardes' paper represents a cautionary story of potentially neuroreductionistic accounts of the subject and human values. She frames the paper by looking at concussion litigation against the National Football League (NFL). Hardes points out that the litigation was based on neuroscientific knowledge, which itself is embedded in a larger social, political, and economic context. Neuroscientific knowledge operates as part of a biological governance system in which health is valorized and risk is viewed as immoral. Hardes warns that certain values attached to risk, such as those in the context of sport, may be threatened.

Jeffrey Fry's 'Two Kinds of Brain Injury in Sport' seeks to broaden the discussion of what constitutes brain injury. Fry argues that pathological emotional conditions following emotional trauma are grounded in brain injury. Thus, brain injury occurs not just as a result of blows to the head or from jarring the brain, but as a result of various forms of human interaction. Given the potential for emotional trauma in sport, he argues that we must broaden the scope of neuroethical concern for brain injury in sport.

Alberto Carrio Sampedro and Jose Luis Perez Trivino's 'On the Compatibility of Brain Enhancement and the Internal Values of Sport' marks a transition from brain injury to brain *enhancement*. The authors consider the compatibility of cognitive enhancement through brain enhancement and the internal values of sport. For illustrative purposes they focus on the drugs methylphenidate (Ritalin) and modafinil (Provigil), and on transcranial stimulation. They conclude that no inherent conflict between cognitive enhancement and the internal values of sport has been demonstrated to date. As a result, they opt for proceeding with caution with further research.

Hub Zwart's 'Skiing and Its Discontents: Assessing the Turist Experience from a Psychoanalytical, a Neuroscientific and a Sport Philosophical Perspective' is a multifaceted discussion drawing on work in the philosophies of film, sport, and empirical investigations—what he calls the 'behavioral scaffold' into the nervous system's responses to anxiety. Using the film 'Turist' as a foil, he compares the skiing resort of the movie as a context for self-knowledge with the laboratory as a context for testing physical and psychic responses to a variety of cues. From this he develops Jean-Paul Sartre's account of skiing as an example of de-technicized experience which represents an urbanization of the sublime.

The next essay explores the hot topic of mirror neurons. Jens Birch's 'Intentional and Skillful Neurons' looks at the role of mirror neurons in action understanding. Birch explores the idea that the motor system is itself cognitive. The upshot is that the mirror neuron system may be intricately involved in the performance of skillful actions in sport, such as those that involve understanding and anticipating competitors' actions. In addition, mirror neurons may provide insights into consciousness that could move forward the debate about the extent to which elite athletes consciously perform athletic tasks.

Andy Fiala's 'Team Spirit, Team Chemistry, and Neuroethics' examines team spirit from a neurobiological and ethical (liberal individualist) perspective. Although Fiala finds much of value in the neurobiological perspective, he cautions against neuroreductionism and argues that we should continue to use ethical, social, cultural, and psychological concepts when analyzing team spirit. Fiala acknowledges that there are both good and bad forms of team spirit and warns of forms of de-individuation, in which the individual is problematically absorbed by the collective. The author concludes with a thought experiment: What if we could artificially induce team spirit by doping the brain? Would this be morally permissible?

In 'High-Level Enactive and Embodied Cognition in Sport', Kevin Krein and Jesus Ilundain-Agurruza adopt an enactivist and embodied approach to cognition. In doing so, they depart from long standing theories in the philosophy of mind wherein mental representation is necessary for cognition. Instead, Krein and Ilundain-Agurruza argue that cognition arises in interaction between agents and environments. While focusing on 'flow' and *mushin* (a concept taken from Japanese martial arts) states, they argue that the enactivist approach helps us understand how high-level cognition is possible in sport. While their approach does not dismiss the importance of the brain, they argue that an enactivist and embodied approach provides a better explanation of skilled athletic performance than does one that relies on mental representations instantiated in neurobiological states. Concomitantly, expert performance helps validate enactivism.

The final essay in the issue, by psychologist Jordan DeLong, is entitled 'Neuropsychology Behind the Plate'. DeLong's paper examines neuro-psychological limitations that affect the accuracy of umpiring. The issues raise questions with both epistemological and ethical import. Since umpires must now reckon with technologies that expose their human frailty behind the plate, should we continue to use flawed, human umpires, or should we dispense with them and leave the 'calls' to be determined by technological devices?

Perhaps unusually for an edited collection, we include an appendix. It comprises an interview with Leonardo Fogassi (one of the world's leading neuroscientists), conducted by Jens Birch, on the subject of mirror neurons. Fogassi was active in the lab at the University of Parma at the time mirror neurons were discovered. This appendix is a prolegomenon to future work on the relevance of mirror neurons to sport. Although the interview is placed at the end of the issue, some readers may find it instructive to begin by reading the interview, and the illuminating prefatory comments by Jens Birch that accompany it.

We hope that these essays are only the beginnings of an ongoing conversation. We would be pleased if this volume spurs further work at the intersection of sport, ethics, and neurophilosophy. Not only are the issues presented here of pressing philosophical and practical concerns, they also represent a new mode of fluid interaction between science and philosophy for the future of sports scholarship.

Notes

1. For the text of Presidential Proclamation 6158 (July 17, 1990), see www.loc.gov/loc/brain/proclaim.html, accessed 9 June, 2017.
2. For an exception to this that goes beyond the debate about concussions, see Fry (2013).

Disclosure statement

No potential conflict of interest was reported by the authors.

References

CORLETT, J.A. 2014. Should inter-collegiate football be eliminated? Assessing the arguments philosophically. *Sport, Ethics and Philosophy* 8 (2): 116–36.

DIXON, N. 2001. Boxing, paternalism, and legal moralism. *Social Theory and Practice* 27 (2): 323–44.

FARAH, M.J. 2013. Neuroscience and neuroethics in the 21st century. In *The Oxford handbook of neuroethics*, edited by Judy Illess and Barbara J. Sahakian. Oxford: Oxford University Press: 761–81.

FRY, J.P. 2013. The neuroethics of coaching. In *The ethics of coaching sports: Moral, social, and legal issues*, edited by Robert L. Simon. Boulder, CO: Westview Press: 151–66.

GLANNON, W. 2011. *Brain, body, and mind: Neuroethics with a human face*. Oxford: Oxford University Press.

LEVY, N. 2007. *Neuroethics*. Cambridge: Cambridge University Press.

PROJECT ON THE DECADE OF THE BRAIN: PRESIDENTIAL PROCLAMATION 6158. (1990, 17 July). Available at www.loc.gov/loc/brain/proclaim.html (accessed 9 June 2017).

RACINE, E. 2010. *Pragmatic neuroethics: Improving treatment and understanding of the mind-brain*. Cambridge, MA: The MIT Press.

ROSKIES, A.. 2002. Neuroethics for the new millennium. *Neuron* 35(3 July): 21–23. doi:10.1016/S0896-6273(02)00763-8.

SAILORS, P.R. 2015. Personal Foul: An evaluation of the moral status of football. *Journal of the Philosophy of Sport* 42 (2): 269–86.

SATEL, S and S.O. LILIENFELD. 2013. *Brainwashed: The seductive appeal of mindless neuroscience*. New York, NY: Basic Books.

Ethics, Brain Injuries, and Sports: Prohibition, Reform, and Prudence

Francisco Javier Lopez Frias and Mike McNamee

ABSTRACT

In this paper, we explore the issue of the elimination of sports, or elements of sports, that present a high risk of brain injury. In particular, we critically examine two elements of Angelo Corlett's and Pam Sailors' arguments for the prohibition of football and Nicholas Dixon's claim for the reformation of boxing to eliminate blows to the head based on (a) the empirical assumption of an essential or causal connection between brain injuries incurred in football and the development of a degenerative brain disease known as chronic traumatic encephalopathy (CTE); and (b) John Stuart Mill's rejection of consensual domination (ie voluntary enslavement). We present four arguments to contest the validity of Corlett, Dixon's and Sailor's positions. Specifically, we argue that (i) certain autonomy-based arguments undermine paternalist arguments for reform; (ii) the nature of the goods people pursue in their lives might justify their foregoing (degrees of) future autonomy; (iii) Mill's argument against consensual domination draws on ambiguous and arbitrary distinctions; (iv) the lack of consensus and empirical evidence regarding CTE arising from brain injuries in sport underdetermines calls for reform. We conclude that these proposals for reforming or eliminating sports with high risks of brain injuries are not well founded.

Might a Ban on Sports with a Risk of Brain Injuries be Justified?

The moral permissibility or impermissibility of those sports in which harms are caused by brain injury is a recurrent theme in sports and medical ethics (Russell 2005, 2007; Breivik 2007; McNamee 2007; Bloodworth 2014; Rosenberg 2014; Simon et al. 2015). The fault lines of these debates were laid down by John Stuart Mill in the middle of the nineteenth century and remain with us today. Thus, in philosophical, ethical, and legal dispute, the notion of vulnerability remains the dominant criterion of demarcation regarding the justified interference on the liberty of individuals pursuing their own life plans and goals. While the debate on the moral permissibility of boxing has spawned its own literature within and beyond the philosophy of sport milieu (Radford 1988) for several decades, the notion of brain injuries in general, and concussion in particular, has raged in public debates, medical science, and in the ethics of sports in recent years.

The ranges of response to such incidents, as have been discussed in relation to American Football, Rugby, and Soccer (Association Football), have been many and varied. They range from uncertainty and denial on the one hand to calls for prohibition on the other. And in the middle ground, there are those who claim—similar to debates in boxing (Parry 1998; Dixon 2001, 2007)—that the merits and internal goods of the activity may be preserved while diminishing its deleterious consequences by a series of reforms. In this article, we discuss two recent contributions to the literature on brain injuries and sports by Corlett (2014) and Sailors (2015) that call for a prohibition on American Football. In addition, we partly draw on Nicholas Dixon's call for reform of boxing, on whose arguments—after John Stuart Mill—Sailors partly relies. We find that the use of Millian arguments are unsound and that the conclusion for prohibition, though it has merit, is not warranted by the arguments offered.

Corlett on the Elimination of Inter-collegiate Football

In his essay 'Should inter-collegiate football be eliminated? Assessing the arguments philosophically', J. Angelo Corlett poses five arguments for the elimination of college football: (a) argument from exploitation; (b) economic argument; (c) academic argument; (d) argument from fraud; and (e) health care and medical costs to others argument. Corlett is not the first to discuss the thorny issues that orbit inter-collegiate athletics (Andre and James 1991; Simon et al. 2015). The nature and status of quasi-professionalized inter-collegiate athletics (especially American Football) within educational institutions are quite unique in the world. Its stadia and merchandising eclipse in commercial and economic terms most professional sports franchises. For this reason, the corruptive character of commercialized inter-collegiate sports, on which Corlett's arguments (a), (b), (c), and (d) rely, has been regarded as one of the main reasons for their elimination. However, according to Corlett, the 'health care and medical costs to others argument' is the only argument that serves its purpose and stands critical analysis not only in the context of college sports but it also in professional football (Corlett 2014, footnote 18). Football, he argues, should be eliminated because:

> it is unfair to saddle others with significant health care and medical costs of, say, CTE [chronic traumatic encephalopathy] given that such costs cannot be realistically covered by [athletes] themselves […] Why should others (taxpayers in general) be forced to cover such costs by way of, say, higher health care insurance premiums? (Corlett 2014, 124)

The central normative concept in Corlett's argument is 'fairness'. For Corlett, athletes and football teams are not able to afford the insurances that provide long-term cover medical and legal CTE and other injuries in the light of new medico-scientific research into the brain and its vulnerabilities. He argues that the only way to cover these programs, in the context of ever-increasing premiums, is through public taxation or higher health care premiums. Since, Corlett argues, football is not a public or social good (a position not without counter-critique) there can be no justifiable reason for forcing people to fund football programs. If individuals—many or most of whom have no desire to support the activity—were obliged to do so, then the insurance covering football would place an unfair financial burden on them. To avoid incurring this unjust situation, football should be eliminated. We note in passing that his argument could be applied to any other sport with a high risk of brain injuries. For example, the Isle of Man TT motorcycle racing has incurred more than 250 deaths in its 110 year history . Nevertheless, it is American Football and Boxing that are the focus of the present essay, being the object of Corlett's, Dixon's and Sailors' arguments.

It is important to note that Corlett's argument relies on some interesting empirical assumptions about the economic structure of sport institutions and nature of sport practices. In particular, Corlett assumes to a large degree the existence and necessary presence of brain injuries in football. He fails to consider seriously the modification of sports, which might allow for a reduction in such injuries either through the introduction of new equipment or the modification of the rules of the game. Earlier, and in a similar vein, Parry (1998) and Nicholas Dixon argue for the prohibition of blows to the head in boxing (Dixon 2001). Moreover, if football were to become economically unsustainable without the support of public funding, the sport itself could be modified to require less expenditure, for example, by reducing team rosters or play time, or by increasing sponsorship, and so on. What Corlett takes seriously is the significance of brain injuries and their consequent economic costs, as opposed to the direct harms to athletes themselves.

Sailors' and Dixon's Millian Argument for the Elimination or Reform of Football and Blow to the Head in Boxing

In contrast to Corlett's case against football, Pam Sailors' and Nicholas Dixon's cases for banning football and blows to the head in boxing, respectively, rely on the same empirical assumption about brain injuries but on a different moral argument, namely, the harm principle (Mill 1989). Corlett dismisses the harm argument as excessively paternalistic. According to him, arguments relying on the medical risks to oneself of playing contact sports like football lead to what Dworkin (1972, 1983) and Feinberg (1984) call 'hard paternalistic justification', that is, 'the attempt to substitute one person's [competent] judgment for another's, to promote the latter's benefit' (Dworkin 1983, 107).

Dworkin's distinction between soft and hard paternalism draws on John Stuart Mill's following famous characterization of autonomy in *On Liberty*:

> The object of this Essay is to assert one very simple principle (…) the sole end for which mankind are warranted, individually or collectively, in interfering with the liberty of action of any of their number, is self-protection. That the only purpose for which power can be rightfully exercised over any member of a civilised community, against his will, is to prevent harm to others. His own good, either physical or moral, is not a sufficient warrant. He cannot rightfully be compelled to do or forbear because it will be better for him to do so, because it will make him happier, because, in the opinions of others, to do so would be wise, or even right. These are good reasons for remonstrating with him, or reasoning with him, or persuading him, or entreating him, but not for compelling him, or visiting him with any evil in case he do otherwise. To justify that, the conduct from which it is desired to deter him must be calculated to produce evil to some one else. The only part of the conduct of any one, for which he is amenable to society, is that which concerns others. In the part which merely concerns himself, his independence is, of right, absolute. Over himself, over his own body and mind, the individual is sovereign. (Mill 1989, 13)

For Mill, autonomy entails the pursuit of 'our own good in our own way, so long as we do not attempt to deprive others of theirs, or impede their efforts to obtain it' (Mill 1989, 16). When individuals possess a tolerable amount of common sense and experience, Mill argues, their way of living is the best because it is *their* own mode of living. Mill's appeal to tolerable levels of common sense and experience opens up a space for rightfully limiting people's sovereignty over themselves. From this standpoint, certain instances of paternalistic policies should be accepted when people are not in the situation to make autonomous choices and would harm themselves involuntarily. Involuntary harm to oneself is likely to occur in situations where agents lack relevant information of the consequences of their actions, are coerced into acting in a certain way, or have a significant degree of mental or physical incompetence. Children and people with disabilities are particularly vulnerable to involuntary harm. Rightful interventions into people's autonomy for their own good based on their lack of autonomy are called 'soft' paternalistic interventions.

Soft paternalism contrasts with hard paternalism. The former protects people from involuntary harm, whereas the latter 'seeks to avert both involuntary and voluntary self-inflicted harms' (Lovett 2008, 127). To illustrate this distinction, Mill uses an example involving an unsafe bridge:

> If either a public officer or any one else saw a person attempting to cross a bridge which had been ascertained to be unsafe, and there were no time to warn him of his danger, they might seize him and turn him back, without any real infringement of his liberty; for liberty consists in doing what one desires, and he does not desire to fall into the river. Nevertheless, when there is not a certainty, but only a danger of mischief, no one but the person himself can judge of the sufficiency of the motive which may prompt him to incur the risk: in this case, therefore, (unless he is a child, or delirious, or in some state of excitement or absorption incompatible with the full use of the reflecting faculty), he ought, I conceive, to be only warned of the danger; not forcibly prevented from exposing himself to it. (Mill 1989, 96)

For Mill, a public officer would only be justified to stop somebody from crossing an unsafe bridge if the bridge-crossing person was not aware of the conditions of the bridge. In case the people attempting to cross the bridge were fully aware of the danger but wanted to do it anyway, the public officer would not be justified in attempting to stop them so doing. On Millian terms—after the previously cited quotation—(s)he might warn them, urge them, remonstrate and reason with them, and so on, but *not* prohibit them. The distinction between soft paternalism and hard paternalism plays a major role in public policy-making in liberal-democratic states. Van der Veer (1986, 13–16) notes forty such instances, including instances such as fluoridization, or compulsory schooling and vaccinations, where we accept such intervention. Most obviously, with regard to smoking cigarettes and drinking alcohol, educative policies are preferred over punitive ones in order to respect the smokers' or drinkers' autonomy. Awareness campaigns to enable citizens to make free choices based on informed decisions about alcohol consumption and tobacco are commonplace, whereas punitive policies are restricted to cases where others are or may be harmed, i.e. the increasing prohibition on smoking in public places, or prohibiting of consumption or purchase by children.

According to Corlett, the situation of the bridge-crossing person is equivalent to that of athletes who choose to play football. Football players might have proper information regarding brain injuries and can use their reflective judgment to knowingly and voluntarily assume the health risks involved. From a Millian perspective, we should provide athletes with accurate and relevant information, which is comprehensible to them, in such a manner to create the conditions for allowing them to make autonomous decisions. As Feinberg argues, Mill's position aligns with the traditional common law maxim *violenti non fit injuria*—to those that consent no injury is done.

Mill's Exception to His Autonomy Principle

Mill, beyond his exceptions of those incapable of autonomous choice, seems to admit a further exception to the application of his principle and view of autonomy as built upon the not-harm-to-others principle, namely, consensual domination (Pettit 2010, 63–64):

> [B]y selling himself for a slave, he abdicates his liberty; he foregoes any future use of it, beyond that single act. He therefore defeats, in his own case, the very purpose which is the justification of allowing him to dispose of himself. He is no longer free; but is thenceforth in a position which has no longer the presumption in its favor, that would be afforded by his voluntarily remaining in it. The principle of freedom cannot require that he should be free not to be free. It is not freedom, to be allowed to alienate his freedom (Mill 1989, 103)

Although Mill argues that we should tolerate people's voluntary decisions to inflict upon themselves harm (even up to and including death, such as the bridge-crossing individual), allowing them to sell themselves into slavery is different (Archard 1990, 546). Anyone who agrees voluntarily to enslavement, according to Mill, is engaged in a performative contradiction. By making the decision of consenting to be dominated by a master, they negate what allows them to act voluntarily. Thus, making a moral choice destroys the very possibility of making such a choice (albeit in the future). Thus, voluntary slavery can be seen in the long run to be self-contradictory. Moreover, if morality is at least in part concerned with the protection and promotion of autonomy, the decision of selling oneself into slavery goes against the key normative principles of morality. This turns the prohibition against voluntary slavery into a rightfully justified restriction.

Dixon calls this kind of restriction 'pre-emptive paternalism', which he defines as the restriction 'on autonomous actions in order to preserve greater future autonomy', and which is well established in our legal system (Dixon 2001, 332). The prohibition on addictive drugs and laws requiring the use of seatbelts and motorcycle helmets are examples of it. Suppose a drink company commercialized an energy drink that caused the impairment of cognitive abilities intrinsically linked with the exercise of autonomy, such as paying attention, organizing thoughts, and remembering. As these effects would surely be detrimental to people's autonomy, impairing their 'ability to reflect on and form values and life plans in the first time, which is the most fundamental element of autonomy' (Dixon 2001, 335), a preemptive intervention would be justified.

Since the effects of the hypothetical energy drink mentioned above correspond to the effects of CTE and other brain injuries, practices leading to them should also be banned. This is Dixon's and Sailor's argument for the elimination of blows to the head in boxing and football, respectively. According to Dixon and Sailors, brain injuries related to practicing football and boxing reduce boxers' and football players' future autonomy the same way selling oneself into slavery does it:

> Given what we know, and are learning, about CTE, choosing to play football is analogous to choosing to be sold into slavery, since choosing football means choosing the likely brain damage that makes later autonomous choice equally impossible. (Sailors 2015, 271).

The practice of sports with high risk of brain injuries, according to Sailors and Dixon, must be seen as analogous to consensual domination. Both types of practices imply the voluntary foregoing of future autonomy. If the reduction of future autonomy provides sufficient grounds for rejecting practices of consensual domination like voluntary slavery, then, it also provides enough grounds for banning activities that allow athletes to trade their future autonomy for the sake of taking part in a sport with a significant risk of brain injuries.

Preemptive paternalism, therefore, justifies the ban on football and blows to the head in boxing. Is this a good argument? Are Dixon and Sailors right? Should we cancel all college football programs and suggest the National Football League (NFL) close its business due to present and prospective effects of brain injuries? Should football and boxing rules and equipment be modified to protect the athlete's future ability to make rational choices?

The following are reasons for doubting that the analogy between voluntary slavery and practicing football and boxing works and, therefore, for accepting Dixon's and Sailor's arguments: (a) it might be that consensual domination is permissible; (b) there might be other present or future compensatory goods that are being overlooked in the argument for eliminating football and blows to head in boxing; (c) Mill's formula is ambiguous and arbitrary, and; (d) the strength of the data regarding the existence of CTE does not support Sailor's and Dixon's use of Mill's slavery analogy.

Four Arguments Against Sailors' and Dixon's use of Mill's Slavery Analogy

The Validity of Mill's Claim that Consensual Domination Should be Forbidden

One possible criticism of Dixon's an Sailors' preemptive paternalistic arguments for the elimination or reform of sports with high risks of brain injuries refers to the validity of Mill's claim that we should not be allowed to sell ourselves into slavery. For example, Dworkin argues that selling themselves into slavery does not preclude individuals from acting autonomously (Dworkin 1983). For him, people should be allowed to forego their future autonomy as long as their decision is the result of their values and rational reflection. Under these conditions, autonomy-limiting decisions can rightfully be made, even if they involve the elimination of future autonomy, on the basis that they fit the life plan of those selling themselves into slavery. People who make such a choice exercise their autonomy by becoming and living as slaves. Dworkin, thus, turns Mill's argument on its head to argue that allowing fully autonomous individuals to sell themselves into slavery should be morally permissible.

Dworkin extends his argument to claim that, not only would forbidding consensual domination limit people's autonomy, it would also impose a particular idea of the good on them (Dworkin 1983). If this is the case, the restrictions on consensual domination should be removed because a society that lets a few people sell themselves into slavery is preferable over one that limits people's autonomy to choose the kind of lives that they want to live. The acceptance of voluntary dominance, thus, would be an 'acceptable price for the sake of preserving liberty' (Archard 1990, 456).

It should be noted that we do not necessarily endorse Dworkin's autonomy driven anti paternalism. We merely point to legitimate limitations in the characterization of preemptive paternalism. Dworkin's criticism of Mill's argument undermines Dixon's and Sailors' use of it to argue for the elimination or reform of sports with high-risk of brain injuries. First, playing football and boxing should be allowed even if their practice results in the elimination of future autonomy. The very choice of engaging in sports like boxing and football is a voluntary one, albeit conditioned by the usual socializing forces and parental influence. Banning the sports would reduce people's autonomy by not allowing them to engage in the kinds of sports that they value. Secondly, Dixon and Sailors should show why a society that limits people's autonomy to practice the sports they like is better than one where individuals can engage in the sports that they find most meaningful. Here Dixon or Sailors might readily look for support in Kant's deontology rather than Millian consequentialism for stronger support. In *On the old saw: that may be right in theory but it won't work in practice*, Kant observes:

> If a government were founded on the principle of benevolence toward the people, as a father 's toward his children—in other words, if it were a paternalistic government (*imperium paternale*) with the subjects, as minors, unable to tell what is truly beneficial or detrimental to them, obliged to wait for the head of the state to judge what should constitute their happiness and be kind enough to desire it also—such a government would be the worst conceivable despotism. (Kant 1793/2006, 290f)

Moreover, who would do the prohibiting and when? Would one keep a register and determine after X concussive incidents that a player would be forbidden from playing? Or would one somehow censure teams whose players seemed to suffer incidences higher than the norm? Or would it follow on lines of legal paternalism that the activity be banned? Here a distinction between boxing or American football might be relevant. There would be little or no danger of the latter going underground so to speak. But the idea that a sport so loved and woven into the fabric of American society and its institutions might be banned on the grounds that the harms attendant on it so threatened the moral fabric of that society would seem a stretch to put it mildly. This would be tantamount of legal moralism (Devlin 2009). And we note that none of the authors resorts specifically to this claim. Nevertheless, it would be incumbent on them to argue who would do the prohibiting and with what kind of authority.

In addition, one would have to consider the ramifications of such prohibition, such as the loss of privacy, opportunity costs of social spending on prosecuting and housing offenders, liberty restrictions on those found guilty, crowded courts and even corruption (Schonsheck 1994). Feinberg (1984) construes these as harms that have to be balanced against the harms that are the object of potential criminalization. Schonsheck (1994) by contrast argues that such a hypothetical test must pass two filters: (i) the costs of criminalization (understood in a broad way as above) must not outweigh the benefits; and (ii) there should be no alternative to criminalization. If we are to prohibit American Football, therefore, on the grounds of harm voluntarily entered into we must also evaluate the indirect consequences of it, and consider the fuller picture that gives potential justification to it. We shall consider this case more fully in section 'How Strong is the Evidence on Which the Sought for Prohibition Exists?'.

Other Goods Argument

Related to the idea that people should be free to choose the kind of lives that they have reasons to value, another criticism of Mill's argument is that it privileges autonomy over other goods that people pursue in their lives. Notwithstanding this, people might value these other goods more than those related to the exercise of autonomy. For instance, let us suppose that a mother gave up her autonomy for a monetary reward to provide for her underfed children. Should this self-sacrificing exercise of autonomy be condemned? This is a very relevant argument in the issue at stake here. Boxers and football players often come from unprivileged minorities, and they find in professional sports a realistic possibility to have a better life for them and their families, or at least to cultivate a range of virtues or excellences that might be both of inherent and instrumental value. Why should they not be allowed to a certain self-sacrifice to provide their loved ones or dependents with better opportunities?

In line with the idea that some goods other than autonomy might be promoted, those relying on Mill's argument should show why the value of autonomy trumps other types of goods or values. Slavery and playing football or boxing are significantly dissimilar when it comes to the valuable goods that they can promote. At the very least, sports provide participants with positive goods like discipline, respect for opponents, self mastery, and so on. In broader views of sport, such as Robert L. Simon's mutualism, sports provide sites for human flourishing by engaging in a voluntarily accepted mutual quest for excellence (Simon et al. 2015). Dixon and Sailors would have to account for these positive goods to utilize Mill's argument against consensual domination to argue for the elimination of autonomy-sacrificing sports. Moreover, the adequacy of the identification between consensual domination and CTE can be contested by appealing to the temporal character of human goods. As Derek Parfit's work on identity and psychological continuity shows, it is rational for individuals to prefer to minimize future suffering and allocate greater suffering in the past (Parfit 2007, §46). In line with this argument, it could be argued that giving up the possibility of enjoying valuable goods in the future in favor of a greater pleasure in the present is justifiable. If this is the case, athletes giving up the possibility of fully employing some future rational capabilities in exchange of more pleasurable lives in the present may be justifiable. Let us elaborate on this point further.

It is widely agreed that prudence should be temporally neutral (Brink 2010). Despite its prevalence in actual decision-making of ordinary people, a short-term bias is typically considered a failure of rationality since one privileges one time slice over another for no other reason than its proximity. Now, a number of qualifications are needed here before we move to a strong critique of the assumptions in Dixon's and Sailor's employment of preemptive paternalism. First, we are not concerned with interpersonal temporal neutrality; only with intrapersonal matters since what Sailors and Dixon, in particular, have argued against is the later loss of autonomy of the injured athlete. Though they recognize the collateral harm that may ensue to family members, for example, it is not the crux of their position. Thus, it is (to use Feinberg's labels) only harms to the self we are concerned with not harms to others. Secondly, even within intrapersonal neutrality, there has been a discussion after Hume (2014) and Parfit (2001, 2007) about the unity of our identity. Challenging as these personal identity theses are, we do not consider them here. Rather our focus is on value conflict over time within the individual. More obviously, the issue is the positive value attached to American football participation in the here and now, and the potential future suffering consequent on it.

In his Felicific Calculus (FC), Jeremy Bentham has put forward the view that propinquity (nearness in time) would be a rational basis on which to prefer one pleasure to another (later) one. This might be seen as impulsive more so than rational (as Sidgwick (2012) observes), so the following qualification needs to be made: where welfare accretions or damages are equal, one may rationally prefer the nearer in time, but this will be on the grounds of certainty (a further principle from the list constituting the FC.[1]

At least at the point of contractual commencement, the Collegiate and NFL player have a relatively clear conception of the benefits at hand. These will be comprised of scholarships and the avoidance of expensive educational fees in the former case (where—contentiously—sponsorships and direct payment are prohibited) and very substantial salaries accrue in the latter case. And, what of the harms? From the standpoint of existing data, we know that football careers are not especially long ones, though the benefits are substantial, more so in the latter case than the former. Yet, we also know that because of recruiting patterns and practices at the high school level, when the lives of young adults are in the balance, that the decision is a precarious one. But precariousness is a weak condition on decision-making when one is considering the outright prohibition of a practice as pervasive and subjectively valued as American Football. This seems sufficient to cast serious doubt on the prohibitionist case. Might more philosophically robust grounds be given to the preference of positive goods in the here and now by the adult athletic reasoner?

Despite commitment to the principle of temporal neutrality, Brink (2010, 356) writes that temporal location might be important when it is considered in contexts where the time of a decision will affect the magnitude of benefits and disbenefits:

> So if at some future point in time I will, for whatever reason, become a more efficient converter of resources into happiness or well-being, however that is conceived, then a neutral concern with all parts of my life will in one sense require giving greater weight to that part of my life. Perhaps, in the 'prime of life' I have greater opportunities or capacities for happiness. If so, temporal neutrality will justify devoting greater resources to the prime of life. However, this is not a pure time preference for that future period over, say, the present, precisely because the same resources yield goods of different magnitudes in the present and the future. The rationality of this sort of discounting is an application of, not a departure from, temporal neutrality. (Brink 2003, 356)

In addition to our vulnerability to the contingencies of life and the existence of moral luck, it strikes us that here is a characterization that is apt for our football playing, brain-injury susceptible, athletes. It is surely uncontentious that at elite levels most sports are time-related goods (Slote 1983). Our capacities for athletic excellence are conditioned by our age and related considerations. To use Aristotelian terms, our potential is not fixed over time. Now in some sports a later maturing profile may be possible, just as other forms of competitive structures that allow for age grading of 'mature' or 'veteran' (in the non-military sense of hat word) may adhere. But these do not apply in American Football as currently constituted. It seems then that the average Collegiate or professional player may be justified in maximizing preferences nearer in time since they greatly outweigh possibilities later in life. And, as Brink observes, this is not *pure* time preference, it is a rational estimation of pains and pleasure from the present point of view with a proper regard for the future insofar as one may estimate it. And this rational choice should be conditioned by a balanced view of the state of CTE evidence alongside other epidemiological studies of injuries arising in the sport.

If it is the case that the preservation of autonomy is concerned, it is more than difficult for the slave analogy to be usefully employed by Dixon or Sailors. The slave analogy draws its power from a form of despotism where autonomy is very severely restricted. Yet, all that can be offered in the context of our potentially brain-injured athletes is that they are making unwise choices, or choices that fail to meet the rational test of temporal neutrality. No one can seriously say that the choices are not *theirs*; that this is not *their* mode of living. And this being the case it seems the best that is open to Sailors and to Corlett and Dixon too, is that a soft paternalism may well be merited with respect to the young or cognitively impaired whose decision-making may be compromised in relation to short and or long term decisions. But it is quite another thing to justify hard or even legal paternalism. In summary, neither Dixon nor Sailors makes a sufficiently robust case for hard paternalism, and the advert to the slavery analogy is grossly distorted.

Mill's Criterion is Ambiguous and Arbitrary

To further show why Sailors' and Dixon's use of Mill's position on consensual domination, on which Sailors' claim for the prohibition of football rests, is significantly problematic we shall focus on the arbitrariness of Mill's argument (Düber 2015; Dworkin 2013).[2] As Strasser (1984) points out, Mill's condemnation of consensual domination is ambiguous and arbitrary since it does not establish a clear-cut point at which the principle of autonomy allows the limitation of people's autonomy. For instance, what is the relevant difference(s) between deciding to cross a damaged bridge risking death, and selling oneself into slavery? Is autonomy not equally eliminated in both cases? In line with this criticism, Sailors and Dixon would have to show what the difference between playing football and blows to the head and the practice of high-risk sports like bungee jumping, wingsuit flying, and BASE jumping lies (Spiegelhalter 2014). For instance, over 11 years and 20,850 jumps, there were 9 deaths and 82 non-fatal accidents among those practising BASE jumping. In light of these data Frank Lovett argues: 'many perfectly ordinary choices that Mill would certainly refuse to subject to social regulation entail reductions in a person's negative liberty' (Lovett 2008, 131). Signing an employment contract, accepting a car loan, purchasing a house, getting married, and having children are voluntary actions that reduce our autonomy. Nobody would argue that the practices like labor, marriage, and owning property should be eliminated for the sake of protecting people's autonomy.

Critics of the use of Mill's analogy to sports that have a predisposition to brain injury could easily find some relevant differences between the loss of autonomy resulting from consensual domination or brain injuries and that caused by other activities like crossing a damaged bridge, getting a car loan, and forming a family. Dixon, for instance, argues: '[These practices] unlike brain damage caused by boxing [do not impair people's] ability to reflect on and form values and life plans' (Dixon 2001, 335). Most of the autonomy-limiting activities presented above are temporally limited or reversible (Lovett 2008). If agents regret having chosen them, they can always do something to recover their lost autonomy (or degree thereof). In line with this, there are several ways to save from arbitrariness and ambiguity Mill's analogy of the bridge-crossing-person example. First, it could be argued that risking one's life is equivalent to the loss of autonomy resulting from selling oneself into slavery. Those who die as a consequence of risking their life do not have the possibility to make free choices anymore. It goes without saying that only those who are alive can. Thus, in high-risk activities, autonomy is not at stake, life is. Secondly, it might be argued that those who engage in a high-risk activity like BASE jumping have no certainty that their autonomy will be reduced. However, those who sell themselves into slavery are not taking a chance; they know with certainty that they will forego their autonomy. BASE jumpers might not lose their life, but football players and boxers only entertain the possibility thereof as a consequence of their brain injuries.

Here is where Corlett's affirmation 'behavioral and brain sciences seem to hold a vital key to the future of football, and perhaps other contact sports' becomes relevant (Corlett 2014, 134). Even if Mill's argument against consensual domination was insurmountable, and the analogy between suffering severe brain injuries and slavery more solid, it remains to be shown that the blows and jolts to the head certainly lead to a severe reduction of the athletes' autonomy. To what extent is this true? Do we have enough scientific evidence to enforce such a claim? We turn now these ethical and scientific concerns now.

How Strong is the Evidence on Which the Sought for Prohibition Exists?

Simon et al. (2015) and Rosenberg (2014) had already noted the lack of causal relationships between significant harms of boxing and mixed martial arts respectively. To respond to the questions posed at the end of the previous section, let us return to the most concise statement of Sailors' position. She writes:

> Given what we know, and are learning, about CTE, choosing to play football is analogous to choosing to be sold into slavery, since choosing football means choosing the likely brain damage that makes later autonomous choice equally impossible. (Sailors 2015, 271)

Let us take the first condition. It is an epistemic one. Sailors begins her case with reference to the claim, published in 1986 in the Journal of the American Medical Association, that American Football is a violent activity in which many injuries are caused. Note that it offers no evidence for what seems a relatively uncontroversial claim. Indeed, the quotation incorporates a claim that there is *no* good evidence 'The peer review literature is apparently mute' (Lundberg 1986 cited in Sailors 2015, 272) and it hints at a conspiracy of the sports medicine profession.

Equally, calls for the banning of boxing and mixed martial arts have appeared with frequency in the British Medical Association's journal, *The British Medical Journal*, which has also published systematic reviews that do not support the widespread beliefs about long-term brain injuries at least in amateur boxing (Loosemore et al. 2008).

Since that time a vast number of studies of varying power and persuasion have appeared. But Sailors does not cite them. Instead, she writes: 'Such studies did not begin to appear for another dozen years, but have become increasingly common as the post-mortem brains of more and more former players have been made available to researchers' (Sailors 2015, 272). What one has to understand here, and which Corlett and Sailor ignore—is that there is deep contentiousness around SRC specifically and brain injury research more generally. Both Corlett and Sailors present empirical viewpoints second hand from a journalistic source, though Corlett is slightly more guarded in his acceptance of their presentation. Yet, journalists like any others attempting to get their point across may not always be even handed with the evidence.

Following Corlett, Sailors' preferred source is journalistic (Fainaru-Wada and Fainaru 2014).[3] These journalists select, from the hundreds of articles available, those of the respected Boston neuroscientists McKee and Cantu. Such selectivity is problematic when one seeks to ban an activity (a) over which the science in its conceptual and empirical dimensions is unclear or incomplete especially with respect to long term effects; and (b) that millions partake in or have partaken in and/or spectate on. A fairer way to proceed, indeed a more analytic way forward, would be to consider a broader range of scientific evidence. This would be typically done by looking at systemic review (e.g. McCrory et al. 2017, which is discussed below), or even metareview articles, that synthesizes research with similar hypotheses and rigorous methodologies in order to bring more powerful data to the table and contrast and compare it meaningfully (Meeuwisse et al. 2017). Certainly, none are as damning or unequivocal as Fainaru-Wada and Fainaru (2014). But before we draw peremptory conclusions there must be clarity at a conceptual level. What clarity and consensus is there regarding sport related concussion (SRC) and CTE?

The counterpart to philosophical, conceptual vagueness within medicine can be called medical epistemic uncertainty. It is widespread in general medicine (Fox 2000) and more so in sports medicine (Malcolm 2009). Given that the causes and conceptual contours of SRC itself is contested we should not be surprised to learn that the concept of CTE is of even greater ambiguity. Moreover, the pathological efforts to describe and diagnose it, as well as efforts to treat it are deeply contested.

Clinicians in sports medicine have tended to be vague about the precise nature of what is referred to as mild traumatic brain injury (mTBI) (McCrory et al. 2013; Slobounov et al. 2013). Signoretti et al. (2010) claim that, 'there are still no standard criteria for the diagnosis and treatment for this peculiar condition'. They outline two alternative approaches to mTBI. The first, which they refer to as the dominant theory, understands mTBI as 'diffuse axonal/neuronal injury' which is divided into the direct mechanical trauma (contusions, hemorrhaging, lesions, lacerations, etc.) and subsequent non-mechanical effects such as swelling, biochemical changes, and so on, that may not develop for hours or even days afterwards. They argue that the lack of lasting neuronal pathology renders this conception weak or at best inconclusive since lasting clinical symptoms are not present 'in the vast majority of patients'.

In contrast to this theory, based on the idea of cell death, is one predicated on the idea of 'temporal neuronal dysfunction, the inevitable consequence of complex biochemical and neurochemical cascade mechanisms directly and immediately triggered by traumatic insult' (Signoretti et al. 2010, 1). They also argue that it is unclear how repeated insults to the brain affects neuronal health (Vagnozzi et al. 2010), though clinicians have worried about this for a long time and refer to it as 'second impact syndrome' (Vagnozzi et al. 2010). Nor is it clear whether, and to what extent, successive blows compound the original injury, nor for what time the brain of the athlete is compromised by the original blow and thus especially vulnerable to further insults to the brain. In a series of articles Vagnozzi et al. (2005, 2008, 2010) even refer to the phenomenon of the 'vulnerable brain'—indicating a window of recovery for up to 30 days until athletes may be symptom-free. This may be especially important since the efficacy of diagnosis may seriously compromise the athlete's present and future health (Vagnozzi et al. 2010). CTE is most likely related to this latter conception of vulnerability.

Now while this brief excursion into brain injury science is doubtless limited from a clinical and scientific viewpoint, it gives clear indication that the existing levels of clinical and scientific confidence. SRC which is seen as a potential precursor to CTE is an emerging area of neuroscience and sports medicine. Until recently, same-day return to play was permitted in American Football (McCrory et al. 2013) which indeed seemed to lack sufficient attention to the duty of care that sports physicians owed their athletic patients, and which seems a proper ethical justification for intervention (McNamee and Partridge 2013; Partridge 2014; McNamee et al. 2015, 2016). Fortunately, the most recent of the 5 global consensus statements (McCrory et al. 2017) has now closed this loophole and recommended the impossibility of same-day return to play for athletes who suffer a concussive episode in a sports event (McCrory et al. 2013).

In the same 2017 Consensus Statement, the authors conducted a systematic review that screened 60,000 articles regarding SRC. Specifically with respect to CTE they report that:

> The potential for developing chronic traumatic encephalopathy (CTE) must be a consideration, as this condition appears to represent a distinct tauopathy with an unknown incidence in athletic populations. A cause-and-effect relationship has not yet been demonstrated between CTE and SRCs or exposure to contact sports. As such, the notion that repeated concussion or subconcussive impacts cause CTE remains unknown. (McCrory et al. 2017, 7)

To be fair, Corlett takes the uncertainty of the evidence more seriously than Sailors though neither fully discounts it. Additionally, it should be noted that the process of appointing expert authors of the Concussion in Sport Group (CISG) has not been without criticism of potential conflicts of interest (Johnson et al. 2015; McNamee et al. 2015). It seems prudent, premised upon this uncertainty, that the precautionary principle be applied, and that players are removed from the field of play following a head injury and an appropriate SRC protocol or test. Thus, following a head injury there are overwhelming reasons to nsist on the removal from the field of play when a head injury occurs despite the subjective reports from the athlete that the feel well enough to carry on. And with children or youth athletes, given the plasticity of their maturing brains there is stronger reasons to mandate the deployment of a child/youth specific protocol or test, and enforce a longer rest period of enforced rest before re-entry following a cessation of symptoms (McCrory et al. 2017)[4]. This soft paternalism is strongly justifiable. Nevertheless, proposing to prohibit the activity en masse on the basis of the journalistic presentation of two respected neuroscientists (among many) in what is a deeply contested scientific field, is more than a step too far. Indeed, it *may* turn out that CTE is caused by current training and performative aspects of sports like American Football but to eliminate or prohibit them entirely is to engage in a paternalism that requires more robust evidence and argument.

What may be justifiable is a kind of 'procedural paternalism' (Beauchamp 2003; Fateh-Moghadam and Gutmann 2014; Vandeveer 2014) that is a form of paternalism that mandates a particular procedure. In recognition of a growing concern, and in the absence of consensus as to the harms caused by brain injuries (notably successive SRCs) temporary intervention may be justified to establish a procedure to establish the conditions of comprehension of the choice to participate. But, again, this falls well short of any justification required to eliminate or prohibit the sport, as Corlett and Sailors argue. That the choice to entertain such risks is the athletes', albeit one conditioned over years of participation, sacrifice, opportunity costs, limitations and influences of various other kinds. Nevertheless, it is a stretch to suggest that the choice to engage in the activities is so wrongful as to be that of engaging in voluntary enslavement or so necessarily harmful as to justify prohibition, and the other personal, economic and social costs that would entail.

Clearly, what is justified at the present time is a significant investment into the education of players, coaches and parents (McNamee et al. 2015, 2016) as well as further basic and applied research into head impacts and jolts. What is also required is a serious consideration of the scientific and medical evidence in a manner where conflicts of interested are mitigated and if possible removed (Partridge 2014). Equally, rule reforms might be considered for American Football such as those that have occurred in Rugby to outlaw tackles that do not involve wrapping the arms of the tackler around the tackled, rather than merely colliding with them. Stricter rules and enforcement of sit out periods (Johnson et al. 2015) and no return to play on the same day as concussive incident (belatedly recognized in the last Concussion Consensus Statement (McCrory et al. 2017) will each add to the harm mitigation of the football player, and those engaged in other sports that create a space for brain injuries, by virtue of the nature of the conflict they enable. Moreover, given the seriousness of the case at hand, a moratorium of kinds, perhaps especially for younger athletes, might be called to allow for these and other revisions to take place.

Conclusion

In this paper, we have criticized one of Corlett's central arguments around the elimination of American Football, and subjected to strong criticism Dixon's and Sailors' claim that sports with high risks of brain injuries should be reformed or eliminated, respectively on the grounds of preemptive paternalism. Dixon and Sailors draw on the Millian moral condemnation of consensual domination. They argue, respectively, that when boxers and football players voluntarily decide to take part in their sports, they give up the very capability that enables them to make such autonomous choices, namely, rational decision-making abilities. According to Sailors and Dixon, by voluntarily giving up their future autonomy, these athletes engage in a performative contradiction and eliminate the very quality—autonomy—that is the ground of their rational choosing. Therefore, since playing football and boxing go against the element upon which people's moral worth is built, the sports should be reformed or banned. Dixon argues for the reform of boxing by eliminating blows to the head. Sailors argues similarly in the case of football. Moreover, since she thinks that reforming football to eliminate the occurrence of CTE is impossible, the entire practice of American Football should be banned.

We have presented four arguments to contest the validity of Dixon's and Sailor's arguments. First, the use of Mill's idea that selling oneself into slavery precludes all future autonomy. An autonomy respectful view, however, would only conclude that human beings should be allowed to pursue the kind of lives that they have reasons to value, even if that involves consensual domination. Secondly, we have argued that the nature of the goods people pursue in their lives might justify foregoing (degrees of) future autonomy. Thirdly, we have shown that Mill's argument against consensual domination draws on ambiguous and arbitrary distinctions. Dixon and Sailors need stronger arguments to ground Mill's distinction for the slavery analogy to be employed decisively. Even if it was proven that Mill's analogy is apt, and that human beings should not be allowed to trade their future freedom for present pleasure, we challenge the empirical assumption that CTE-related injuries are equivalent to consensual domination. To do so, fourthly, we have highlighted the lack of consensus and empirical evidence regarding the long term SRC consequences and their relationship to CTE. We conclude that the proposals from reforming or eliminating sports with high risks of brain injuries are at present not well founded.

Notes

1. Of the FC specifically, one might further consider the fecundity of the activity, though we do not do that here.
2. Corlett's position is less damaged by this problem and is more circumspect. He writes: '… even if it turns out that medical and brain sciences reveal that CTE is caused by football and other high-contact sports, it is unclear that a paternalistic argument counts as a sufficiently good reason to prevent informed and consenting adults from participation therein' (Corlett 2014, 124). Thus, on hard paternalistic grounds, Corlett neither rules prohibition in or out. Nor does he offer reasons why he equivocates. However, he does argue that soft paternalistic interventions may be merited over children, on Millian grounds, but he gives no details as to how this might be operationalized. On whether this would take the form of a legal intervention, with the problems we have averted to, or merely rule based reforms (or any other option in between), he is silent.
3. On a point of accuracy, Corlett cites one more journalistic source: Farrar (2012) but that hardly strengthens his case.

4. The protocols have developed over the years of the Consensus statements including more heightened protections for children and youths, those these were always more strict than for adults. See McCrory (2017) for details of the relevant protocols.

Acknowledgements

Thanks to Jeff Fry and Andrew Edgar for their comments on earlier versions of this paper.

Disclosure Statement

No potential conflict of interest was reported by the authors.

References

ANDRE, J., and D.N. JAMES. 1991. *Rethinking college athletics*. Philadelphia, PA: Temple University Press.
ARCHARD, D. 1990. Freedom not to be free: The case of the slavery contract in J. S. Mill's on liberty. *The Philosophical Quarterly* 40 (161): 453–465. doi:10.2307/2220110
BEAUCHAMP, T.L. 2003. Methods and principles in biomedical ethics. *Journal of Medical Ethics* 29 (5): 269–274. doi:10.1136/jme.29.5.269
BLOODWORTH, A. 2014. Prudence, well-being and sport. *Sport, Ethics and Philosophy* 8 (2): 191–202. doi: 10.1080/17511321.2014.935741
BREIVIK, G. 2007. The quest for excitement and the safe society. In *Philosophy, risk, and adventure sports*, edited by M. McNamee. London: Routledge: 10–25.
BRINK, D.O. 2003. Prudence and authenticity: Intrapersonal conflicts of value. *Philosophical Review* 112 (2): 215–245.
BRINK, D.O. 2010. Prospects for temporal neutrality. In *Oxford handbook of the philosophy of time*, edited by C. Callender. Oxford: Oxford University Press: 353–381.
CORLETT, J.A. 2014. Should inter-collegiate football be eliminated? Assessing the arguments philosophically. *Sport, Ethics and Philosophy* 8 (2): 116–136. doi:10.1080/17511321.2014.918167
DEVLIN, P. 2009. *The enforcement of morals*. Indianapolis, IN: Liberty Fund.
DIXON, N. 2001. Boxing, paternalism, and legal moralism. *Social Theory and Practice* 27 (2): 323–344. doi:10.5840/soctheorpract200127215
DIXON, N. 2007. Sport, parental autonomy, and children's right to an open future. *Journal of the Philosophy of Sport* 34 (2): 147–159. doi:10.1080/00948705.2007.9714718
DÜBER, D. 2015. The concept of paternalism. In *New perspectives on paternalism and health care*, edited by T. Schramme. Dordrecht: Springer International Publishing: 31–45. doi:10.1007/978-3-319-17960-5_3
DWORKIN, G. 1972. Paternalism. *The Monist*, 64–84.
DWORKIN, G. 1983. Paternalism: Some second thoughts. In *Paternalism*, edited by R.E. Sartorius. Minneapolis: University of Minnesota Press: 105–112.
DWORKIN, G. 2013. Defining paternalism. In *Paternalism: Theory and practice*, edited by C. Coons and M. Weber. New York, NY: Cambridge University Press: 25–38.
FAINARU-WADA, M., and S. FAINARU. 2014. *League of denial: The NFL, concussions, and the battle for truth*. New York, NY: Three Rivers Press.
FARRAR, D. 2012. James Harrison says that new helmet padding protects him after 'double-digit concussions.' *Shutdown Corner*, 10 July. Available at https://www.yahoo.com/blogs/nfl-shutdown-corner/james-harrison-says-helmet-padding-protects-him-double-172643262–nfl.html
FATEH-MOGHADAM, B., and T. GUTMANN. 2014. Governing [through] Autonomy. The Moral and Legal Limits of "Soft Paternalism". *Ethical Theory and Moral Practice* 17 (3): 383–397. doi:10.1007/s10677-013-9450-3
FEINBERG, J. 1984. *Harm to others*. Oxford: Oxford University Press.
FOX, R. C. 2000. Medical uncertainty revisited. In *Handbook of social studies in health and medicine*, edited by G. Albrecht, R. Fitzpatrick, and S. Scrimshaw. United Kingdom: SAGE: 409–425. doi:10.4135/9781848608412.n26

HUME, D. 2014. *An enquiry concerning the principles of morals*. Oxford: Oxford University Press.
JOHNSON, L.S.M., B. PARTRIDGE, and F. GILBERT. 2015. Framing the debate: Concussion and mild traumatic brain injury. *Neuroethics* 8 (1): 1–4. doi:10.1007/s12152-015-9233-8
KANT, I. 2006. *On the old saw: That may be right in theory but it won't work in practice*. Philadelphia: University of Pennsylvania Press ; University Presses Marketing [distributor].
LOOSEMORE, M., C.H. KNOWLES, and G.P. WHYTE. 2008. Amateur boxing and risk of chronic traumatic brain injury: Systematic review of observational studies. *British Journal of Sports Medicine* 42 (11): 564–567.
LOVETT, F. 2008. Mill on consensual domination. *Ten* 123–137.
LUNDBERG, G.D. 1986. Boxing should be banned in civilized countries—Round 3. *JAMA* 255 (18): 2483–2485. doi:10.1001/jama.1986.03370180109044
MALCOLM, D. 2009. Medical uncertainty and clinician-athlete relations: The management of concussion injuries in rugby union. *Sociology of Sport Journal* 26 (2): 191–210. doi:10.1123/ssj.26.2.191
MCCRORY, P., MEEUWISSE, W.H., AUBRY, M., CANTU, B., DVOŘÁK, J., ECHEMENDIA, R.J., … TURNER, M. 2013. Consensus statement on concussion in sport: The 4th International Conference on Concussion in Sport held in Zurich, November 2012. *British Journal of Sports Medicine* 47 (5): 250–258. doi:10.1136/bjsports-2013-092313
MCCRORY, P., W. MEEUWISSE, J. DVORAK, M. AUBRY, J. BAILES, S. BROGLIO, … P.E. VOS. 2017. Consensus statement on concussion in sport—the 5th international conference on concussion in sport held in Berlin, October 2016. *British Journal of Sports Medicine* 51: 838–847.
MCNAMEE, M. 2007. Adventurous activity, prudent planners and risk. In *Philosophy, risk and adventure sports*, edited by M. McNamee. London: Routledge: 1–10.
MCNAMEE, M.J., and PARTRIDGE, B. 2013. Concussion in sports medicine ethics: Policy, epistemic and ethical problems. *The American Journal of Bioethics* 13 (10): 15–17. doi:10.1080/15265161.2013.828123
MCNAMEE, M.J., PARTRIDGE, B., and ANDERSON, L. 2015. Concussion in sport: Conceptual and ethical issues. *Kinesiology Review* 4 (2): 190–202. doi:10.1123/kr.2015-0011
MCNAMEE, M.J., PARTRIDGE, B., and ANDERSON, L. 2016. Concussion ethics and sports medicine. *Clinics in Sports Medicine* 35 (2): 257–267. doi:10.1016/j.csm.2015.10.008
MEEUWISSE, W.H., SCHNEIDER, K.J., DVORAK, J., OMU, O.T., FINCH, C.F., HAYDEN, K.A., and MCCRORY, P. 2017. The Berlin 2016 process: A summary of methodology for the 5th International Consensus Conference on Concussion in Sport. *British Journal of Sports Medicine* 51: 873–876
MILL, J. S. 1989. *On liberty: With the subjection of women and chapters on socialism*. Edited by S. Collini. Cambridge: Cambridge University Press.
PARFIT, D. 2001. *On what matters*. Oxford: Oxford University Press.
PARFIT, D. 2007. *Reasons and persons*. Oxford [u.a.: Clarendon Press.
PARRY, J. 1998. Violence and aggression in contemporary sport. In *Ethics and sport*, edited by M. McNamee and J. Parry. London & New York: E & Fn Spon: 205–224.
PARTRIDGE, B. 2014. Dazed and confused: Sports medicine, conflicts of interest, and concussion management. *Journal of Bioethical Inquiry* 11 (1): 65–74. doi:10.1007/s11673-013-9491-2
PETTIT, P. 2010. *Republicanism: A theory of freedom and government*. Oxford [8][u. [9]a.: Oxford University Press.
RADFORD, C. 1988. Utilitarianism and the noble art. *Philosophy* 63 (243): 63–81.
ROSENBERG, D. 2014. Violence in sport. In *The bloomsbury companion to the philosophy of sport*, edited by C.R. Torres. New York, NY: Bloomsbury Publishing: 262–275.
RUSSELL, J. S. 2005. The value of dangerous sport. *Journal of the Philosophy of Sport,* 32 (1): 1–19. doi:10.1080/00948705.2005.9714667
RUSSELL, J. S. 2007. Children and dangerous sport and recreation. *Journal of the Philosophy of Sport,* 34 (2): 176–193. doi:10.1080/00948705.2007.9714720
SAILORS, P. R. 2015. Personal foul: An evaluation of the moral status of football. *Journal of the Philosophy of Sport* 42 (2): 269–286. doi:10.1080/00948705.2014.1000338
SCHONSHECK, J. 1994. *On criminalization – An essay in the philosophy of criminal law*. Dordrecht: Kluwer Academic Publishers.
SIDGWICK, H. 2012. *The methods of ethics*. Cambridge: Cambridge University Press.
SIGNORETTI, S., VAGNOZZI, R., TAVAZZI, B., and LAZZARINO, G. 2010. Biochemical and neurochemical sequelae following mild traumatic brain injury: summary of experimental data and clinical implications. *Neurosurgical Focus* 29 (5): E1. doi:10.3171/2010.9.FOCUS10183

SIMON, R.L., P.F. HAGER, and C.R. TORRES. 2015. *Fair play: The ethics of sport*. Boulder: Westview.
SLOBOUNOV, S. M., TEEL, E., & NEWELL, K. M. 2013. Modulation of cortical activity in response to visually induced postural perturbation: Combined VR and EEG study. *Neuroscience Letters* 547: 6–9. doi:10.1016/j.neulet.2013.05.001
SLOTE, M.A. 1983. *Goods and virtues*. Oxford: Clarendon Pr.
SPIEGELHALTER, D. 2014. Extreme sports: What are the risks? *BBC.Com*.
STRASSER, M. 1984. Mill and the utility of liberty. *The Philosophical Quarterly* 34 (134): 63–68.
VAGNOZZI, R., SIGNORETTI, S., TAVAZZI, B., CIMATTI, M., AMORINI, A. M., DONZELLI, S., … LAZZARINO, G. 2005. Hypothesis of the postconcussive vulnerable brain: Experimental evidence of its metabolic occurrence. *Neurosurgery* 57 (1): 164–171; discussion 164–171.
VAGNOZZI, R., SIGNORETTI, S., TAVAZZI, B., FLORIS, R., LUDOVICI, A., MARZIALI, S., … LAZZARINO, G. 2008. Temporal window of metabolic brain vulnerability to concussion: A pilot 1H-magnetic resonance spectroscopic study in concussed athletes–part III. *Neurosurgery* 62 (6): 1286–1296; discussion 1295–1296. doi:10.1227/01.neu.0000333300.34189.74
VAGNOZZI, R., SIGNORETTI, S., CRISTOFORI, L., ALESSANDRINI, F., FLORIS, R., ISGRÒ, E., … LAZZARINO, G. 2010. Assessment of metabolic brain damage and recovery following mild traumatic brain injury: A multicentre, proton magnetic resonance spectroscopic study in concussed patients. *Brain: A Journal of Neurology* 133 (11): 3232–3242. doi:10.1093/brain/awq200
VAN DER VEER, D. 1986. *Paternalistic Interventions*. Princeton, NJ: Princeton University Press.
VANDEVEER, D. 2014. *Paternalistic intervention: The moral bounds on benevolence*. Princeton, NJ: Princeton University Press. Available at http://public.eblib.com/choice/publicfullrecord.aspx?p=1701173

Governing sporting brains: concussion, neuroscience, and the biopolitical regulation of sport

Jennifer Hardes

ABSTRACT
Drawing on the recent concussion litigation from the United States' National Football League (NFL), the paper examines the emergence of neuroscience knowledge as part of a defining rationale for the justification and rationalization of the lawsuit. The paper argues that neuroscience knowledge is best understood as a regulatory discourse that is attached to larger social, political, and economic realities that bring it into being as a legitimate type of knowledge. This larger socio-political governance logic is one that scholars call 'biopolitical' which emphasizes the protection of individual life over and above other ways of being. Risk discourses that frame risk-taking practices as immoral thus emerge within this biopolitical regime of governance that frame morality in terms of public health that individual citizens ought to pursue. With this in mind neuroscience knowledge plays an important role in concussion litigation. It emerges as a technology of biopolitical governance in that it is used to justify legal decisions on concussion. This is despite the fact that neuroscience knowledge remains nascent and even scientifically uncertain. Because of this, the paper argues that scholars ought to not only consider neuroscience research skeptically, but also ought to be aware of the dangers of neuroscience's emergence as an 'anticipatory discourse' that has the potential to reduce human behavior to matters of the brain that thus transforms our very ontology of ourselves and the practices we perceive as 'good'.

Introduction

In July 2011 the Superior Court of California, Los Angeles County heard a suit filed by 75 former professional National Football League (NFL) players against the NFL and helmet company, Riddell Inc. (*Maxwell et al. v National Football League et al., (Superior Court of California, County of Los Angeles, BC, 19 July 2011*). The retired players claimed that NFL Parties had not only put players at risk from concussive and sub-concussive injuries but that the NFL had fraudulently concealed these risks. Three successive lawsuits were filed that were then consolidated and joined by approximately a further 5000 former players who have since filed over 300 lawsuits against the NFL. Former players claimed that the NFL not only deceitfully concealed risks from repetitive traumatic brain injury but that the NFL also cultivated and 'glorified a culture of violence and a gladiator mentality, encouraging NFL players to play despite head injuries' (re: National Football Player's Concussion Injury Litigation [2015], MDL 2323, p. 6). These series of lawsuits were eventually turned into a class action lawsuit in 2014–2015 that was concluded in 2016 when NFL players Keith Turner and Shawn Wooden, on behalf of themselves and other similarly situated plaintiffs, filed to hold the NFL responsible for alleged injuries based on liability and breached duty of care. Resisting these claims, the NFL responded that 'there was no scientifically proven link between repetitive traumatic head impacts and later-in-life cognitive/brain injury' (2015, Id. 308). Yet, silently acknowledged at the centre of this case as that which was encumbered with speculation and scientific uncertainty, but that would be used to tenaciously support the weight of the case, was a pivotal type of knowledge: neuroscience.

It is this focus on neuroscience and its role in the construction of knowledge claims about concussion injuries in sport law that is central to this paper. In particular, the paper is concerned with the knowledge claims that neuroscience makes to 'truth' and how this truth gets used as a regulatory mechanism in reorienting the 'goods' of sport toward what I describe as biopolitical governance agendas preoccupied with 'life itself' at the expense of other value orientations. With this in mind, the paper opens with a brief overview of the rise of neuroscience knowledge and its claims to normativity before next explaining the knowledge claims central to the concussion litigation and the role of neuroscience knowledge in legitimating these claims. Next, the paper moves onto considering how the legitimation of concussion litigation through neuroscience knowledge can be understood within a larger understanding of neuroscience as a regulatory discourse that is part of a broader biopolitical mode of governance. Finally, the paper considers how such a regulatory discourse shifts, and indeed puts into question, the very 'goods' of sport itself.

We are our synapses

Since it emerged as a professional field of inquiry in the 1960s, neuroscience knowledge has become increasingly dominant in our mainstream understanding of the human being and, particularly, the mind. The 1990s, hailed as the 'decade of the brain', revealed the weight of neuroscience as a field and discipline of knowledge (Satel and Lilienfeld 2013). Questions about human consciousness that have baffled psychologists, physiologists, and philosophers for decades appeared to be concluded through the one organ in the human body that was suddenly being revealed to us, not only by neuroscientists themselves, but by their host of sophisticated imaging technologies that could precipitously provide a window to our souls and reach the depths of human consciousness. Approximately 100 billion neurons in the brain that interact through the trillion synaptic connections have come to be regarded as the key to unlocking humanity and providing the answer to every element of human behavior (Herculano-Houzel 2009; Mahfoud 2014).

Scholars like Stephen Morse and Nikolas Rose describe this incursion of neuroscience knowledge as 'neuroexuberance' (Morse 2011) and 'neuromania' (Rose and Abi-Rached 2013). While they do not deny the importance of neuroscience knowledge per se, they are concerned with the rapid shift toward attempting to understand and reduce the human subject to the brain that has exploded both in science, in fields beyond neuroscience itself, and in popular media. As neuroscience knowledge has become accepted as a mode of enhancement and diagnosis, concerns emerge that neuroscience claims to challenge and open up new conceptualizations of society, of subjects, and of the human brain that reshape our fundamental ontologies of ourselves (Rose 2007). Speculation emerges from these scholars not least because the 'explanatory gap' that has kept philosophers, psychologists and physiologists speculating for centuries appears to have been filled with a new 'materialist ontology of thought' that has not, as Rose (2016) claims, been discerned by philosophers, but rather by technology, and human interpretation of this technology. A new ability to 'see the mind'—or see the mind simply as the brain—causes unease for those who seek to conceive of the human in less reductionist fashions and to also critically appraise and caution the normative force of such neuroscience discourses that take root in our understanding of humanity.

Scholars dealing with this question of the normative force of neuroscience knowledge have also been increasingly concerned with its role in the regulation and governance of life. Speculations arise about where such new conceptualizations of the human subject framed around the brain may take us in our understanding of the very function of what it means to be human, how we relate to one another, and to what extent we are able to function as free willing subjects. These concerns are coupled with 'harder' concerns regarding the regulation of neuroscience knowledge through the legal realm (Morse 2015a, 2015b; Schleim 2012). There is a concern that the law—and indeed sport law—may take on neuroscience knowledge in such a way that it changes the very foundations of law itself and the fundamental principles that underscore it (Green and Cohen 2004).

We have already seen evidence of the permutations of neuroscience knowledge as they make their way into sport. Where the majority of studies are performance enhancing, neuroscience knowledge is also used for diagnostic and rehabilitative, as well as regulatory, purposes, as evident in the opening examples from this paper of the NFL cases. Thus, in sport, in law, and in daily life, the rise of neuroscience knowledge as a legitimate mode of enhancement and as a diagnostic rationale reveals itself as an increasingly dominant way to understand and explain human behavior.

Concussion, neuroscience knowledge, and scientific uncertainty

Whilst concussion was identified in the tenth century it continues to be a relatively unknown injury or illness in modern medicine, primarily because of its 'invisibility' (Stone, Patel, and Bailes 2014). Generally, concussion is regarded as a head injury that results in a temporary loss of normal brain function, differentiated from other head injuries by its effects on consciousness. Early on, concussion symptoms were analogized to alcohol affliction, termed 'punch drunk'. At the turn of the twentieth-century X-ray technology, that had only recently become available, made possible concussion diagnosis as the presence of a skull fracture or cerebral contusion (Stone, Patel, and Bailes 2014). Punch drunk syndrome was also linked to 'dementia pugilistica', regarded as long-term effects of repeated, low-grade concussive brain injuries. In 1954, medical researchers identified these chronic symptoms in a neuropathological report on a retired boxer who had post-traumatic dementia and Alzheimer's Disease symptoms (Stone, Patel, and Bailes 2014). Studies on boxing continued in the 1970s but it was only in 2005 that a study by neuropathologist Bennet Omalu and colleagues emerged on the case of NFL player, Mike Webster, that sought to identify a relation between concussion and dementia pugilistica, coined at that time Chronic Traumatic Encephalopathy (CTE).

Omalu's study helped the diagnosis of concussion, previously made possible through visual markers and cognitive tests, become more scientifically determinable through the association between concussion and CTE that could be revealed through postmortem and neuropathological examinations of the athlete's brain. Postmortem examination procedures involve the examination of the brain under a microscope, revealing that repeated trauma leads to a build-up of an abnormal tau protein. Tau protein is said to be responsible for stabilizing nerve cell structure; when abnormal it forms twisted fibers in the nerve cells, destroying them which can cause neurological diseases. Brain imaging technologies have promised to provide a deeper scientific insight into concussion and concussion-related illnesses through the rise of PET scans in the late 1950s and fMRIs in the 1990s (e.g. Barrio et al. 2016), though at present these scans have yet to identify CTE in living athletes and only postmortem examinations are claimed to be conclusive (Saulle and Greenwald 2012).

Despite imaging technologies having not managed to diagnose CTE in living athletes, the NFL litigation was hinged on the development of neuroscience knowledge that could make this link conclusive. Players in the NFL class action lawsuit alleged that their engagement in football, through which they experienced concussive and sub-concussive injuries, was the cause of various medical conditions linked to CTE such as Alzheimer's Disease, Dementia, depression, cognitive functioning deficits, processing speed reduction, memory loss, inattention and reasoning, sleeplessness, mood swings, and personality changes. Players filing the lawsuit claimed that the NFL was aware of these 'scientifically proven links' between concussive and sub-concussive injuries and CTE markers of tau protein but did not educate NFL players of the dangers; such dangers, plaintiffs argued, were clearly evident given that in 1994 the NFL Parties formed a Mild Traumatic Brain Injury Committee (MTBI Committee). Players, therefore, accused the NFL of negligence and fraud, claiming that the NFL had knowledge of concussion related injuries in these earlier cases (pre action lawsuit).

In response, objectors claimed that links between concussion and CTE were not scientifically credible and remained uncertain. Since Omalu's research only 200 brains had been examined for CTE. A link between concussion and CTE could therefore not be conclusively stated. This scientific uncertainty of CTE was not concealed by the judge: the case fully acknowledged the 'nascent' state of CTE knowledge and that '… the symptoms of the disease, if any, are unknown' (*Re: NFL 2015*, p. 79). As the case clearly stated, '[I]t should be emphasized that an association is not equivalent to causation' (82). Despite the clear scientific uncertainty, neuroscience knowledge prevailed as a legitimate rationale for the decision held against the NFL.

Scientific uncertainty, risk and the biopolitics of sport

Scientific uncertainty in law is a common point of contention. In cases of public controversy, such as the issue of concussion, scientific uncertainty becomes inexplicitly linked and '… intertwined with differences in policy and philosophy' (Weiss 2003, 28). Karl Popper had claimed that science could only ever approximate to truth, but as Weiss (2003) acknowledges, this approximation is also not politically or philosophically neutral.

One of the major concerns, if not the main organizing principle, of the concussion litigation case was with regard to public health and the risk that players, future players, and society more broadly faced, from cultivating a sporting culture that valued the 'gladiatorial' discourse of pain and injury. It was this point raised forcefully by players who believed they had been inculcated into a culture of risk-taking practices that had normalized pain and injury and therefore normalized concussion as part of the sport itself. Players felt this normalization of concussion was made possible through a concealment of research that indicated the contrary: that indeed concussion was a serious long-term health problem.

Litigation from players is particularly interesting given that, as scholars like Parissa Safai (2003) have claimed, sporting practices generally do indeed '… normalize pain and injury in sport' (127) in a 'culture of risk' that pushes players '… to accept risk-taking … and to make light of the consequences of injuries' (128). One could argue that players have adopted this culture of risk somewhat freely. While one could argue that sports like football do breed risk taking and violence and, therefore, NFL Parties are responsible for inculcating this environment, particularly with regard to younger players, it is equally fair to say that athletes have agency to resist these norms; to think otherwise would be to accuse athletes of being cultural dupes.

Where Safai has noted a culture of risk, she also notes a 'culture of precaution' with regard to the rise of the public health agenda concerning health and safety more broadly speaking. In this culture of precaution risk-taking is regarded as an immoral value (Beck 1992; Giddens 1999). Such a shift to a politico-philosophy that decouples risk, pain, and injury from the 'goods' of sport seems evident in the NFL concussion litigation, in contrast to the existence of sport as an exceptional realm that validated character traits of valiance, courage and risk taking; risky practices associated with hegemonic masculine norms and celebrations of 'gladiatorial' brutality are being othered. Far from being role models, players are instead medicalized and their engagement questioned; their injuries suffered and illnesses diagnosed in later life are thus inextricably linked to a failure of sports' morality and a failure of the practices of governance—in this instance the NFL Parties—to protect players from this risk.

Taking responsibility for one's health and well-being is such a newly founded valiant, courageous and noble act. Athletes engaging in responsible health promotion activities are one means of 'becoming moral'. Sport, then, has been decoupled from an instrumental rationality of risk that licenses certain risky practices that have an instrumental good attached to the state, such as those risks engaged with by public sector personnel, or private sector personnel who benefit members of public, such as the military, police, and fire services who face daily health risks.

Arguably this shift in the normative weight of sport practices like football can be understood as part of a larger biopolitical shift in governance toward practices of self-care and health management. Rose and Abi-Rached (2013), for example, speak of neuroscience knowledge as a new cog in the wheel of the biopolitical management of life that seeks to express the value of life and the ontology of being in an economy of health. Neuroscience values brain health and well-being as part of this wider health economy where the 'good life' values health over and above all else. Aristotle's distinction between *zoe* and *bios*, or bare and political life, has been blurred (Agamben 1998). The rise of the 'social' realm or 'matters of the household' as Arendt (1958) had noted, emerges as *the* central political good. For scholars like Foucault (1990, 2008) and Italian philosopher, Esposito (2008) biopolitics transpired as an outcome of a liberal, and increasingly neoliberal, political rationality of governance that focuses on the protection of individual human life; for Esposito these individual protective measures around life emerge at the expense of more collective forms of social relations, and close off other ways of being that do not conform to valuing life itself above and beyond other commitments and drives. Part of this rationality that seeks to govern and manage the unit of individual human life at the level of the population is a strategy that seeks to individualize and accord each subject responsibility for his or her own 'life'; governing individuals' 'souls' as Rose (1990) puts it, turns individuals' health into a moral issue and ensures subjects monitor their health and thus their soul as moral subjects accordingly.

While being careful not to demonize sport given its association with public health goods (i.e. exercise and health promotion), sporting practices like football that engage athletes in risky cultures, it seems, can no longer be exceptionalized without attention to how they might mitigate such health concerns. Indeed, even in exceptional political spheres that do engender rationalized 'appropriate' risks for the larger public good as noted, risk mitigation and health promotion strategies are becoming increasingly apparent in outward facing health promotion strategies (e.g. Chiarelli 2010; Ranby et al. 2011).

Discourses of risk and anticipation

As noted with reference to legal decision-making and settling lawsuits, framing, managing, and regulating any kind of risk such as concussion is also not politically neutral: institutional and political actors have vested interests in constructing and framing risk in relation to sporting concussion through 'laws, technological interventions and rule changes' (Bachynski and Goldberg 2014, 324). Where risky corporeal practices are decoupled from the good life, risk is also used as a concept to engage the larger public health rationale of governance. This arises through what Pickersgill (2011) calls an 'anticipatory discourse'.

The notion of anticipatory discourses overlaps with the nature of 'predictive and preventive medicine' (Rose and Abi-Rached 2013) that have a 'promissory character' (Williams, Katz, and Martin 2012, 241); they anticipate what might emerge and put regulatory measures in place to deal with these problems if and when they come to fruition. Anticipatory discourses and risk are interlinked through what Ulrich Beck and Anthony Giddens refer to as a 'risk society' that is central to modernity and is a society '… in which we increasingly live on a high technological frontier which absolutely no one completely understands and which generates a diversity of possible futures' (Giddens 1999, 3). This risk society is closely associated with anticipatory discourses because risk is a problem of modern society that is preoccupied with the future.

Law and neuroscience are both anticipatory discourses in the sense that they do not simply respond and decide on things said, but instead actively construct future actions. In the case of the latter, neuroscience knowledge operates as a rationality of governance that engenders future actions by promising something other than what is. In the case of concussion this arises by way of future oriented explanations through the framing of 'neurofutures' (Martin 2015; Williams, Katz, and Martin 2012, 241). The legal case of *Re: NFL Players* (2016) makes this point clear: the case was hinged on the future promise of diagnostics. Where neuroscientists agree that CTE has only been diagnosed posthumously, researchers describe the 'holy grail' of CTE research as that which can diagnose CTE in living players. The case further argued that, despite research being nascent, it was clear that knowledge was building in the direction of the case; an argument was thus made that it would be unfair to discriminate against players because knowledge had not yet confirmed a link between concussion and CTE, and it was generally accepted that the tentative link was strong 'enough'. For instance, it was stated that 'the settlement recognizes that knowledge about CTE will expand

… Arguably, these uncertainties exist because clinical study of CTE is in its infancy' (79). Later: 'the settlement has some mechanism for keeping pace with science, in that the parties must meet and confer every ten years in good faith about possible modifications to the definitions of Qualifying Diagnoses' (90). From these examples one can glean that scientific and public expectations mobilized in and through law also help mobilize neuroscience knowledge as legitimate; in turn, neuroscience knowledge also works to legitimate law as a forward-thinking institution that is not simply dogmatic and prescriptive but that instead responds to public concerns.

The problem with anticipatory discourses is not simply the content of their claims but the function their engagement has on the field of neuroscience knowledge more broadly and on the governance of subjects more specifically. For Pickersgill (2011) such discourses add to the 'neuromania' and 'neuroexuberance' cautioned by Rose and Abi-Rached and by Morse noted earlier, in turn defining neuroscience as knowledge that must be taken seriously, bolstering its normative weight in the regulation of practices like sport. This normative weight is important to consider with reference to the very nature of the normative: neuroscience knowledge is framed in the language of scientific progress such that diagnosis will eventually be made entirely possible and will inevitably reveal the truth of what is claimed in the law. Thus, the law appeals to a futurity to rationalize its decision based on technological and scientific progress. Anticipation, risk, uncertainty, and diagnosis thus go hand in hand such that 'risk' can be realized and framed around wider governance agendas (Simon 2005).

Diagnosis, materialities, and brain imaging technologies

Despite anticipatory discourses being ambiguous they cannot survive on thin air and instead require the circulation of different scientific 'evidences' as a weight of support and rationale behind them. Pickersgill (2011) argues that imaging technologies are one such route through which this is made possible. Being regarded as an 'invisible injury' was helpful for the rationalization of decisions on concussion, because imaging technologies claimed to make it visible—or at least to make what is alleged to be the result of concussion, CTE, visible, thus adding credibility to neuroscience knowledge. While the knowledge produced by technology remains tentative and ambiguous, the interpretation and take up of the knowledge is transformable because it allows us to 'see' and imagine risk through visual biomarkers.

Emerging in various forms from X-rays invented in 1895 by the German physicist William Conrad Roentgen to computerized axial tomography scans, into more 'advanced' technology with PET and fMRI, neuro diagnostics have been presented as more reliable, scientific and objective (Satel and Lilienfeld 2013). In cases of concussion, imaging technologies such as MRIs have been used to examine brain anatomy, specifically grey matter, to view areas of the brain that appear to have 'scarring' from concussive related incidences, and more recently fMRIs, used to examine brain activity, have examined how concussive brains have 'abnormal blood flow' (Leddy et al. 2013). While experts have claimed that these imaging technologies do not currently display accepted evidence of CTE, they are nonetheless posited as technologies and biomarkers that can, and will, reveal the truth of this invisible injury. Neuroscience knowledge has thus changed the very ontology of concussion, providing scientists and clinical practitioners with what emerges as new 'evidence'. The 'explanatory gap' is filled with a new 'materialist ontology of thought' that has not, as Rose (2016) claims, been discerned by philosophers, but rather with technology, that has made visible the mysterious and invisible.

For anthropologist, Joseph Dumit, imaging technology builds '… assumptions into its architecture and thus can appear to confirm them, while … reinforcing them' (2004, 81). Imaging technologies produce truths and confirm these truths through diagnoses that present themselves as objective factual data, despite the data they produce being highly interpretive. A central point raised in Satel and Lilienfeld's critical text *Brainwashed* focuses on the importance of 'interpretation' in neuroscience knowledge that is often presented factually. Examining neurocorrelates on a brain scan is not value free science but rather is highly interpretive. That the reliability of brain imaging technologies and their interpretations have been questioned by neuroscientists themselves ought to be part of such a speculative analysis; a publication by Vul et al. (2009) highlighted this point, noting 'puzzlingly high correlations' that imaging studies claim to reveal.

Knowledge production concerning the materiality of the human mind now understood as brain also allows one to unify visual, material 'results' and diagnose effectively. Diagnostics have 'unifying consequences' in that they don't just 'sort things out, they also link things together' (Rose 2013, 4). Sociologists interested in diagnostics consider not simply what the diagnosis of an illness or injury is, but also the diagnosis' effects on subjects and on society, as well as how social values and subjects shape the very need and desire for such diagnostics. Petersen and Lupton's (1996) remarks on the 'new public health' suggest that diagnostics give people control in a time when the prospect of disease and illness is deemed high.

As a tool for diagnostics, imaging technologies are part of neurosciences' regime of normativity (Pickersgill 2011). The medicalization of concussion knowledge makes possible diagnostic claims that in turn produce new regulatory practices. Through media that displays brain scans of the normal and abnormal CTE brain, neuroscience knowledge permeates and is accepted as truth. The seductiveness of imaging technologies for the public and for those involved in the medical and legal realm also ought then to become a point of contention: the fMRI '… confers a great deal of scientific credibility to studies of cognition and that these images are one of the primary reasons for public interest in fMRI research' (McCabe and Castel 2008, 344). The general public and the media, as well as those engaged in the legal profession, can be seduced into the claims that areas of the brain that are either normal or deviant regarding certain functions can be identified and corrected, and can likewise be seduced into regarding certain practices as being more or less moral.[1] The 'discovery' of CTE through neuropathological examination alongside neuroscientific means of imaging technologies that link concussion to diseases of the brain provide the 'proof' needed to civilize sport in line with larger public health agendas.

Diagnosis does not simply have a role in unifying disease but also has a role in making disease social; it is 'an apparatus through which individuals 'make up' themselves and one another' in order to differentiate between '"normal" and "pathological" experiences … relate to substances and practices … and negotiate situations saturated with moral feelings and implications' (Pickersgill 2014, 521). By unifying the consequences of concussive-related injuries, CTE as its final formulation manifest in the brain, can be made known, diagnosed, and the practices related to it—in this case sport—can be made subject to moral evaluation. Those participating in these practices can also be made subject to moral speculation and assessment.

In contrast to these skeptical views, one might argue that we should embrace these diagnostics rather than criticize them. Neuroregulatory mechanisms that make concussive injury abnormal rather than part of the normal hegemonic sport culture liberates players from brutality and violence; it also detaches players from abnormal behavior and puts them into a context where they can be diagnosed, worked on, and rehabilitated into normal society. Diagnosis is thus 'transformative' as it allows individuals the chance to embark on a 'moral career' that enables them to change their self perception (Goffman 1958; Rose 2013).

This transformative potential made possible through neuroscience knowledge is particularly significant for NFL players who have received various degrees of media attention regarding violent and aggressive tendencies off the pitch as well as on it; spectacular headlines regarding suicides of retired players, instances of domestic abuse, as well as murder suicide in the case of Joseph Belcher have engulfed media.[2] Medical diagnostics provide a possible rationale for this behavior that sideline the individual athlete's responsibility, placing the blame instead on the constitutive rules of the sport. Understood in the wider discourse of public health, 'sport as courageous' is transformed into 'sport as dangerous': for the individual engaged in this transformation of society's morality of sport, his or her behavior can be understood as a byproduct of a more fundamental moral problem with the practices inculcated in sport.

It is interesting to note here that the NFL class action lawsuit anticipated this relation between morality and behavior, and that this association was carefully dealt with, maintaining some distance between a complete reduction of the player's morality to sport-related concussion injuries. In doing so the case drew a distinction between neurocognitive symptoms of CTE and 'mood and behavioural' symptoms that might be regarded as 'deviant'. Where objectors claimed that CTE could be linked to both neurocognitive symptoms and mood and behavioral symptoms, the settlement drew a clear line between these arguing that the former had to be differentiated from the latter because the latter '… are common in the general population and have multifactor causation' (86), '… such as exposure to major lifestyle changes, a history of drug or alcohol abuse, and a high Body Mass Index ("BMI")' (41).

Those objecting claimed that such distinctions ought not to be made because mood and behavioral symptoms are precursors to neurocognitive ones. For example, athletes suffering from CTE are alleged to go through stages of headaches, depression, suicidality, and aggression before experiencing the neurocognitive symptoms. Yet such distinction and categorization was important for two reasons. First, the distinction enabled the legitimation of imaging technologies in rationalizing concussion as a 'real issue' based on its relation to CTE. This is because neurocognitive symptoms found in tau protein, for example, could arguably be traced to 'real' findings in the brain. Neuroselves could be seen in these images. In contrast it would have been regarded as more speculative to propose a link between a disease and behavior that could not so easily be 'validated'; this is despite attempts being made to 'find' and 'locate' behavior in specific regions of the brain.

Second, the law likely did not want to rely on a reductionist view of the subject that attributed behavioral symptoms to concussion-related injury because governance relies on the very notion that subjects can work on themselves and become good moral citizens in line with social norms. Morse (2015a, 2015b) argues that neuroscience knowledge will not change fundamental concepts of law such as freedom, culpability, and responsibility and that such claims are simply 'brain overclaim syndrome'. However, it is not simply the case that the law could be fundamentally undone, as he argues, by neuroscientific reductionism of human behavior to the brain: rather it is also because governance requires we are constructed as free-willing subjects responsible for our own behavior. One must have the possibility to be rehabilitated and to work on one's self in order to become a good, moral citizen. Differentiating between behavioral and neurocognitive symptoms ensured medicalization was possible while still leaving space for the cultivation of responsible citizens who could choose to engage in good moral practices.

Conclusion

As medical ethicist and sociologist Faulkner (2003) writes, 'Evidence and governance are closely linked' in the field of health care. Health care is regulatory and credentialist as well as scientific. This applies also to neuroscience. It, too, is regulatory, credentialist, as well as scientific and thus discussions of neuroscience knowledge in sport ought not to be conceptualized, ethically, without this broader socio-political view. This paper has argued that when links between CTE and sport concussions are shored up through legal proceedings that draw heavily on neuroscience knowledge, concussion becomes not simply an issue for individual athletes, or for the NFL or other sporting bodies, but rather it is reframed as a public health concern. The medicalization of concussion and CTE made possible through technological windows to the previously unseen soul, reframes concussion, but more importantly reframes the sport of football more broadly construed, as a social problem. As a result, the reformulation of the values of life around classic liberal paternal views of risk and the benefits of the protection of individual life are at stake. Health—both of the individual athlete and of the broader public health—becomes the ultimate good that law and regulatory practices must protect, despite views to the contrary expressed by many.

This is not to suggest that health ought not to be part of a reflexive morality, particularly for those of us involved in, and who value, sport. This paper has not suggested that neuroscience knowledge is good or bad per se; rather, neuroscience knowledge is problematic when it becomes a regulatory discourse that is not critically analyzed in line with larger governance agendas that attempt to regulate and orient individuals toward a view of the good life that is, itself, subject to normative governance agendas. Indeed, in many respects, one could argue that neuroscience knowledge in the instance of concussion and football is a positive normative force: in response to a culture of sport that celebrates violence and glorifies a gladiatorial mentality, neuroscience knowledge could potentially be transformative in undoing the dominant hegemonic practices embedded in that culture that exceptionalize the practice of sport violence. On the other hand, however, we ought to also consider the practices that regulatory discourses close off. Moral connotations and expectations about how to live the good life and how to be a good citizen are attached to instrumental risk discourses (Lupton 1993) that, framed around public health, open up these existential questions regarding the ultimate good of human life. Perhaps this is a relative or particular view of ethics, but it is important to critically the wider socio-political context in which normative appeals to health are claimed, recognizing that not, in every instance, do subjects choose health over and above other values. Subjects engaging in practices normatively constructed as 'unhealthy' might not be deemed 'immoral' subjects as they are typically tarnished, but rather as subjects who value other ways of being that are not defined by an instrumental view of the health and longevity of life (Blencowe 2012; Metzl and

Kirkland 2010). Ethicists ought to question whether health and the rationalization of the protection of the subject ought to take such normative centre stage (Conrad 1994).

In short, neuroscientific knowledge ought to be ethically and critically reflected on as it is: discursive and regulatory, attached to larger systems of governance and constructs of the good. As a regulatory science we must be cautious of neuroreductionism: that is, the reduction in human experience to the brain, but also the reduction in the ultimate moral 'good' to 'brain health' and even more broadly, athlete health, that can be discerned through scientific evidence. Neuroscience knowledge is used to diagnose, unify and import such moral decisionism on sport; it therefore ought not to be closed off to discussion but rather should remain an open point of debate for those involved in sport ethics. How does neuroscience as a discourse help reorient the good life toward the larger governance agenda of public health? What is the good of sport participation if it is not *simply* biopolitical? Is there a moral good of sport, and for that matter a moral good of life in general, beyond 'life itself'?

Notes

1. Racine, Bar-Ilan, and Illes (2005) for example coined the term 'neurorealism' that claims that brain images are more real than other forms of data on human behavior (Satel and Lilienfeld 2013, 21). Imaging technologies allow us to view mental disorders as 'real things' thus contributing to their legitimation.
2. Dr Piotr Kozlowski, the physician preparing the postmortem pathology report claimed Belcher likely suffered from CTE given findings of tau protein tangles in seven sections of the brain hippocampus.

Disclosure statement

No potential conflict of interest was reported by the authors.

References

AGAMBEN, G. 1998. *Homo sacer: Sovereign power and bare life*. Translated by Daniel Heller-Roazen. Stanford, CA: Stanford University Press.
ARENDT, H. 1958. *The human condition*. Chicago, IL: University of Chicago Press.
BACHYNSKI, K. and D. GOLDBERG. 2014. Youth sports & public health: Framing risks of mild traumatic brain injury in American football and ice hockey. *The Journal of Law, Medicine & Ethics* 42(3): 323–333.
BARRIO, J., G. SMALL, K.-P. WONG, S.-J. HUANG, J. LIU, D. MERRILL, C. GIZA, R. FITZSIMMONS, B. OMALU, J. BAILES, and V. KEPE. 2016. In vivo characterization of chronic traumatic encephalopathy using [F-18]FDDNP PET brain imaging. *Proceedings of the National Academy of Sciences in the United States of America*.
BECK, U. 1992. *Risk society: Towards a new modernity*. London: Sage.
BLENCOWE, C. 2012. *Biopolitical experience: Foucault, power and positive critique*. Basingstoke: Palgrave Macmillan.
CHIARELLI, P. 2010. *Army health promotion, risk reduction, suicide prevention, report*. Darby, PA: Diane Publishing.
CONRAD, P. 1994. Wellness as virtue: Morality and the pursuit of health. *Culture, Medicine and Psychiatry* 18 (3): 385–401.
DUMIT, J. 2004. *Picturing personhood: brain scans and biomedical identity*. Princeton, NJ: Princeton University Press.
ESPOSITO, R. 2008. *Bios: Biopolitics and Philosophy*. Minneapolis, MN: University of Minnesota Press.

FAULKNER, A. 2003. In the sociomedical laboratory of citizen health: Exploring science, technology, governance and engagement in prostate cancer detection in the UK. *Cardiff Institute of Society, Health & Ethics Working Paper*. Available at http://www.cardiff.ac.uk/socsi/resources/wp74.pdf.

FOUCAULT, M. 1990. *The history of sexuality, volume 1: An introduction*. New York, NY: Random House.

FOUCAULT, M. 2008. *The birth of biopolitics: Lectures at the College de France, 1978–79*. Translated by G. Burchell. New York, NY: Palgrave Macmillan.

GIDDENS, A. 1999. Risk and responsibility. *Modern Law Review* 62 (1): 1–10.

GOFFMAN, E. 1958. *The presentation of self in everyday life*. Edinburgh: University of Edinburgh Social Sciences Research Centre.

GREEN, J. and J. COHEN. 2004. For the law, neuroscience changes nothing and everything. *Philosophical Transactions for the Royal Society of London Biological Sciences*. 359 (1451): 1775–1785.

HERCULANO-HOUZEL, S. 2009. The human brain in numbers: A linearly scaled-up primate brain. *Frontiers in Human Neuroscience* 3(31): 1–11.

LEDDY, J., J. COX, J. BAKER, D. WACK, D. PENDERGAST, R. ZIVADINOV, and B. WILLER. 2013. Exercise treatment for postconcussion syndrome: A pilot study of changes in functional magnetic resonance imaging activation, physiology and symptoms. *Journal of Head Trauma Rehabilitation* 28 (4): 241–249.

LUPTON, D. 1993. Risk as moral danger: The social and political functions of risk discourse in public health. *International Journal of Health Services* 23 (3): 425–435.

MAHFOUD, T. 2014. Extending the mind: A review of ethnographies of neuroscience practice. *Frontiers in Human Neuroscience* 8(359): 1–9.

MARTIN, P. 2015. Commercialising neurofutures: Promissory economies, value creation and the making of a new industry. *Biosocieties* 10 (4): 422–443.

MCCABE, D. and A. CASTEL. 2008. Seeing is believing: The effect of brain images on judgments of scientific reasoning. *Cognition* 107: 343–352.

METZL, J. and A. KIRKLAND. 2010. *Against health: How health became the new morality*. New York, NY: NYU Press.

MORSE, S. 2011. The status of neurolaw: A plea for current modesty and future cautious optimism. *The Journal of Psychiatry & Law* 39: 595–626.

MORSE, S. 2015a. Neuroscience, free will, and criminal responsibility. Edited by Walter Glannon. *Free will and the Brain: Neuroscientific, philosophical, and legal perspectives* (Cambridge 2015); University of Pennsylvania Law School, Public Law Research Paper No. 15-35. Available at SSRN http://ssrn.com/abstract=2700328.

MORSE, S. 2015b. Neuroprediction: New technology, old problems. *Bioethica Forum* 8 (4): 128–129.

PETERSEN, A. and D. LUPTON. 1996. *The new public health: Health and self in the age of risk*. London: Sage.

PICKERSGILL, M. 2011. Connecting neuroscience and law: Anticipatory discourses and the role of sociotechnical imaginaries. *New Genetics & Society* 30 (1): 27–40.

PICKERSGILL, M. 2014. Debating DSM-5: Diagnosis and the sociology of critique. *Journal of Medical Ethics* 40 (8): 521–525.

RACINE, E., O. BAR-ILAN, and J. ILLES. 2005. Science and Society: fMRI in the public eye. *Nature Reviews, Neuroscience* 6 (2): 159–164.

RANBY, K., D. MACKINNON, A. FAIRCHILD, D. ELLIOT, K. KUEHL, and L. GOLDBERG. 2011. The PHLAME (Promoting Healthy Lifestyles: Alternative Models' Effects) firefighter study: Testing mediating mechanisms. *Journal of Occupational Health Psychology* 16 (4): 501–513.

ROSE, N. 1990. *Governing the soul: Shaping the private self*. London: Free Association Books.

ROSE, N. 2007. *The politics of life itself: Biomedicine, power and subjectivity in the twenty first century*. Princeton, NJ: Princeton University Press.

ROSE, N. 2013. *What is diagnosis for?* Institute of Psychiatry Conference on DSM-5 and the Future of Diagnosis. Available at http://nikolasrose.com/wp-content/uploads/2013/07/Rose-2013-What-is-diagnosis-for-IoP-revised-July-2013.pdf (accessed 4 June 2016).

ROSE, N. 2016. Neuroscience and the future for mental health? *Epidemiology & Psychiatric Sciences* 25 (2): 95–100.

ROSE, N. and J. ABI-RACHED. 2013. *Neuro: The new brain sciences and the management of the mind*. Princeton, NJ: Princeton University Press.

SAFAI, P. 2003. Healing the body in the 'culture of risk': Examining the notion of treatment between sports medicine clinicians and injured athletes in Canadian intercollegiate sport. *Sociology of Sport Journal* 20: 127–146.

SATEL, S. and S. LILIENFELD. 2013. *Brainwashed: The seductive appeal of mindless neuroscience*. New York, NY: Basic Books.

SAULLE, M. and B. GREENWALD. 2012. Chronic traumatic encephalopathy: A review. *Rehabilitation Research and Practice*. 1–9. Article ID: 816069. doi:http://dx.doi.org/10.1155/2012/816069.

SCHLEIM, S. 2012. Brains in the context of the neurolaw debate: The examples of free will and 'dangerous' brains. *International Journal of Law and Psychiatry* 35: 104–111.

SIMON, J. 2005. Risk and reflexivity: What socio-legal studies add to the study of risk and the law. *Alabama Law Review* 57 (1): 119–139.

STONE, J., V. PATEL, and J. BAILES. 2014. The history of neurosurgical treatment of sports concussion. *Neurosurgery* 75 (4): S3–S23.

VUL, E., C. HARRIS, P. WINKIELMAN, and H. PASHLER. 2009. Puzzlingly high correlations in fMRI studies of emotion, personality, and social cognition. *Perspectives on Psychological Science* 4 (3): 274–290.

WEISS, C. 2003. Expressing scientific uncertainty. *Law, Probability and Risk* 2: 25–46.

WILLIAMS, S., S. KATZ, and P. MARTIN. 2012. Neuroscience and medicalization: Sociological reflections on memory, medicine and the brain. In M. Pickersgill & I. Van Keulen. *Sociological perspectives on neuroscience*. Bingley: Emerald Group Publishing: 231–254.

Two Kinds of Brain Injury in Sport

Jeffrey P. Fry

ABSTRACT
After years of skepticism and denials regarding the significance of concussions in sport, the issue is now front and center. This is fitting, given that the impact of concussions in sport is profound. Thus, it is with trepidation that one ventures to direct some attention onto brain injuries other than concussions incurred through sport. Given a closer look, however, it may be that considering various kinds of brain injuries, with different causes, will help us better understand the range and seriousness of brain injuries in sport. My focus is on emotional trauma in sport. I argue that severe emotional pathology is evidence of underlying brain injury. Any attempt to minimize the significance of emotional disturbance that results from emotional trauma as 'merely' psychological in nature is thus misguided.

1. Introduction

After years of skepticism and denials regarding the significance of concussions in sport, the issue is now in the spotlight. As evidence accumulates, it is becoming clear that the impact of concussions is profound.[1] Nevertheless, the medical lexicon still sends mixed messages, as evidenced by the use of the term 'mild traumatic brain injury' (MTBI) to designate a class of concussions.[2] Given this context, it is with trepidation that one ventures to redirect some spotlight onto possible brain injuries in sport involving factors other than concussions. However, by broadening the scope of our concern, perhaps we will come to a better understanding of the range and seriousness of brain injuries, and of the contributing factors.

There is something in particular about the brain that perhaps helps account for why brain injuries in sport have suffered relative neglect. In his book *The Absent Body* (1990), Drew Leder writes:

> Unlike the body surface, visible to self and Other, the brain rarely makes an appearance in the life-world. Admittedly, it is available in certain limited ways; we see brains at autopsy, through diagnostic imaging techniques, or pictured in textbooks… Yet in the everyday life-world it is highly unusual to encounter a brain. (Leder 1990, 111)

Elsewhere Leder writes of 'a seeming paradox'.

> That organ which subtends embodied consciousness is itself unavailable to conscious apprehension… It… recedes from my direct apprehension even as it grounds my experience. (Leder 1990, 113)

In this paper, I thematize the typically absent brain, and, in particular, the brain in an injured condition. But brain injuries need not be caused by external blows to the head or by otherwise violently jarring the brain. Instead of focusing on concussions, I highlight contribution1s to brain injuries by problematic transmissions in our interrelationships, or at what psychologist Louis Cozolino (2014, xiv–xv) calls the 'social synapse'. With this emphasis in mind, I argue that emotional disturbances that are the aftermath of emotional trauma sustained in various ways in the context of sport are evidence of gradations of underlying brain injury.[3] Perhaps some individuals who suffer from such emotional disturbances can be found among what Catherine Malabou (2012, 10) calls the 'new wounded'. By the 'new wounded', Malabou (2012) refers to 'every patient in a *state of shock* who, without having suffered brain lesions, has seen his or her neuronal organization and psychic equilibrium permanently changed by trauma'.

Following LeDoux (2015, 18–21), by 'emotions' I refer to 'conscious feelings'. Again following LeDoux (2015, 19–20), I also have in view nonconscious processes that 'contribute to emotional feelings', but which LeDoux says 'should *not* be labeled as "emotional"'. Thus the brain is able to detect and initiate a 'nonconscious response to danger'. This can lead to further brain activity that results in a conscious experience.[4]

This inquiry implicates overlapping metaphysical, conceptual, empirical, and normative issues. With respect to metaphysics, the age-old mind–body problem resurfaces, as well as the recently posed issue of 'brain-body dualism' (Glannon 2013, 14). A related issue concerns how we should conceptualize the nature of injury, and its relationship to harm. A central empirical matter pertains to how we identify or locate a brain injury. Finally, taking seriously the embrained condition of the athlete prompts neuroethical questions. In sum, we may need to broaden our understanding of what constitutes injury, what injury 'looks' like (a key issue in this essay), and how injury occurs and may be detected. In addition, a broadened consideration of brain injury may elicit further ethical reflection on sporting practices, and, in some cases, a call to reform.

In what follows, I first consider the nature of injury and how vestiges of dualism cloud our understanding of when the designation 'brain injury' is applicable. Second, I examine varieties of brain injury, some with subtler properties than others, and I nominate emotional trauma as a cause of brain injury. Third, I turn to emotional trauma in sport and its ethical significance. In a fourth and final section, I respond to some potential objections to my claims.

2. What is an Injury?

In his book *Harm to Others* (1987), the philosopher Joel Feinberg posits a distinction between harm and injury. Feinberg writes:

> Harming in the relevant sense is sometimes contrasted and sometimes identified with injuring, but this confusion is easily eliminable. We are apt to think of injury as specific damage done to the body—broken bones, lacerations, 'internal injuries.' So conceived, injury is one kind of harm. In ordinary speech, persons are not said to be injured by inflictions of harm to interests other than that in physical health and bodily integrity, except by analogy. Psychological shock, for example, is a kind of harm analogous to physical traumas, and can be called an 'injury,' though not without some linguistic strain. (A physician attending a dazed soldier might find it more natural to say that 'he has not been injured, only shocked.') The more distant the analogy to physical wounds, the less appropriate the term 'injury.' (Feinberg 1987, 106)

In Feinberg's account of ordinary language, the term 'injury' designates a kind of harm and applies primarily to 'physical wounds'. 'Psychological shock' is also considered a form of harm, but it is only injury in an analogous or derivative sense of the word 'injury'. These purported distinctions in ordinary language create a bifurcation that resonates with certain forms of mind–body dualism.[5]

One encounters varieties of dualism in the philosophy of mind. René Descartes (1993) famously posits a *strong* form of mind–body dualism. His position, commonly referred to as substance dualism, states that the mind and the body are separate substances with different kinds of properties. In particular, minds are immaterial, while bodies are material in nature. At the same time, Descartes subscribes to the view that minds and bodies interact. Herein lies a problem that Descartes bequeaths to the history of philosophy. Given the purported differences between minds and bodies, how does a Cartesian dualist intelligibly explain the nature of their putative interaction? Indeed, how does one establish that there is a reliable connection between what goes on in minds, on the one hand, and bodies, on the other? If one accepts the Cartesian account of the mind–body relationship, then it *does* seem difficult to establish a close, noncoincidental correlation between psychological shock and physical injury.

But the Cartesian account obscures possibilities that could closely link psychological shock to physical injury. Consider a token identity version of physicalism. According to one such view, each token mental state is identical to some token brain state or other. So, a state of psychological shock, conscious or not, would be identical to a brain state. For an alternative account, consider property dualism, with its more robust ontology. For a certain kind of property dualist, psychological shock, insofar as it involves a conscious mental state, might be considered a property that is supervenient upon and realized in a state of the brain (Heil 2013, 183–198; Kim 2011, 122–125).[6] If either token identity theory or property dualism as presented here is correct, then human psychology is grounded in physicality, and even by the ordinary speech criteria posited by Feinberg, psychological shock would implicate brain injury, by either being identified with or caused by a neurological state. Even some versions of substance dualism might posit an indirect connection between mental states and brain states. For example, theologians who hold a view called psychophysical parallelism might posit that when a psychological state occurs a correlative state of the brain also occurs, not because the psychological state and the brain state are identical or causally related, but because God pre-established that they would occur in tandem.

Thus, in various theoretical accounts emotional disturbance could, in either a weaker or stronger sense, be connected to brain injury. My own intuition is that the connection is a strong one. Yet even if this intuition is well-founded perplexing questions loom. Consider again Feinberg's notion of 'psychological shock', which is an admittedly vague notion. Suppose that psychological shock implicates brain injury in a strong sense, as an identity theorist or a certain kind of property dualist might hold. Is then psychological shock or a kindred state a necessary and/or sufficient condition for positing brain injury? Is brain dysfunction a prerequisite of brain injury? This line of inquiry leads to a consideration of the varieties of brain injuries, and of the diverse methods of detecting them.[7]

3. What is a Brain Injury?

During the nineteenth century, French physician Paul Broca conducted a postmortem exam of a patient called 'Tan', so-named because he frequently uttered the sound 'tan'. In the course of his exam, Broca discovered brain lesions, visible to the naked eye, that established a link between an area of the left frontal lobe (so-called Broca's area) and speech production (Finger 1994, 36–38).

More recently, as attention has turned to Alzheimer's disease, magnification and staining techniques have revealed the hallmark presence of beta-amyloid plaques and neurofibrillary tangles in brains of people with the disease. As authors Mark Fainaru-Wada and Steve Fainaru discuss in their book *League of Denial: The NFL, Concussions, and the Battle for the Truth* (2013), neurofibrillary tangles of tau protein without clear indication of Alzheimer's disease have also been found during postmortem exams of the brains of former National Football League (NFL) players, raising questions about possible connections between their football careers and their brain pathologies.

But not all brain abnormalities follow these paradigms. With the advent of modern neuroimaging technologies, such as magnetic resonance imaging (MRI) and functional magnetic resonance imaging (fMRI), it is now not only possible to detect gross lesions, tumors, or other histopathologies, but also the size of brain structures, and patterns of neurological activity. These new technologies as well as future advances may help us detect brain injuries with novel and subtle kinds of pathological properties that don't fit preconceived notions of what injury 'looks' like.[8] Malabou (2012) writes in a kindred vein about a broad conception of 'brain damage' in connection with trauma victims:

> The behavior of subjects who are victims of trauma … display striking resemblances with subjects who have suffered brain damage. It is possible to name these traumas 'sociopolitical traumas' … Today, however, the border that separates organic trauma and sociopolitical trauma is increasingly porous.
>
> This affirmation tends to generalize and enlarge the concept of brain damage opening it to types of harm that do not initially pertain to neuropathology. It is thus necessary to show that all trauma impacts neuronal organization, particularly the sites of emotional inductors. This is precisely the point that makes it possible to construct a paradigm for all other 'new' wounded. In addition, this affirmation makes it possible to understand neuronal disturbance in other terms than pure and simple physiological lesions. (Malabou 2012, 10–11)

Today's mantra is that the brain exhibits plasticity.[9] The brain is responsive. For example, through learning one's neurological profile changes. Of particular significance for our purposes is the fact that the brain is responsive to our interactions with others. Cozolino (2014, 389) writes that 'we are individuals, but the architecture of our brains also constitutes records of our interpersonal histories'. This raises a complex and important question. When is brain change pathological and when is the brain merely exhibiting its plastic nature?[10]

If we follow Feinberg (1987) suggestion that injury is a kind of harm, then brain injury entails harm to the brain. But does brain injury require a noticeable cognitive deficit, emotional disturbance (conscious or otherwise) or other functional impairment? Suppose that the brain has redundant systems so that loss of viable neurons has negligible effect, or that the brain, through its plasticity, is able to rewire following a traumatic event, such as a stroke.[11] Or consider the fact some children affected by Rasmussen's syndrome have even been able to adapt remarkably well to the removal of one hemisphere of the brain. In cases like these, does the brain remain in an injured condition? One approach would be to say that the previously injured brain has recovered from the injury, though it exists now without previously functioning parts. I find that such cases pose a conundrum, and my intuitions are conflicted regarding whether a detectable dysfunction (psychological or otherwise) is a *necessary* condition for brain injury. However, I *am* arguing that where there is a significant and sustained cognitive deficit or emotional disturbance, this is *sufficient* evidence to posit both psychological harm and injury to the brain, even if we are unable at the present time to pinpoint the specific mode and location of the injury, such as by identifying neural correlates of the condition. Some conditions may require difficult judgment calls as to whether brain injury is present, but other cases seem clearer, such as severe cases of (posttraumatic stress disorder) PTSD.[12] Among the classic symptoms of PTSD is that of disturbed memory systems.

Much has been learned about the relationship between memory in its various guises and the brain. Memory can be compromised by various means, not just in cases involving PTSD. One of the most famous cases in the history of neuroscience, involving H.M., or Henry Molaison (died 2008), involved anterograde amnesia (the inability to form new long-term memories) after the surgical removal of medial temporal lobe structures. H.M. retained memory of distantly past events and procedural memory, evidenced by the fact that he was able to master new learning tasks, though he could not remember that he had engaged in learning trials. Among the things learned from H.M. was that there are different kinds of memory, some of which continue to function as other capabilities are lost (Corkin 2013).

In another famous case, from the early twentieth century, French physician Edouard Clarapede treated a female patient with anterograde amnesia. Each time he met her he would reintroduce himself since the patient could not remember their previous encounters. On one occasion he greeted her with a tack in his palm, so that when they shook hands her hand was pricked. On a subsequent visit she refused to shake hands, although she could not explain why. Later this behavior would be interpreted as evidence for the distinction between an explicit memory system, on the one hand, and an implicit emotional memory system, on the other (LeDoux 1998, 180–182).

In addition to its role in implicit emotional memory, however, emotion colors explicit memory. Emotional arousal affects memory processing (in either an enhancing or inhibitory way) at the encoding, consolidation and retrieval stages (Johnston and Olson 2015, 183–192). This is relevant to the discussion of PTSD and to emotional trauma in general. Emotional trauma is, on the surface, somewhat paradoxical in that it can lead to both loss of certain memories and yet selectively heighten memory of other events. In particular, PTSD is noted for recurring flashback experiences (van der Kolk 2014, 66–67). In his book, *The Body Keeps the Score: Brain, Mind, and the Body in the Healing of Trauma* (2014), Bessel van der Kolk, medical director of a trauma center, and a professor of psychiatry at Boston University, writes of how trauma can cause 'people to become hopelessly stuck in the past' (van der Kolk 2014, 10). People who have experienced trauma project their trauma onto present circumstances. Their mental flexibility is compromised (van der Kolk 2014, 17). This condition is, I posit, reflected in the brain. Neuroscientists speak of looking for the engrams or storage mechanisms for memories in the brain.[13] These are brain alterations that reflect learning. Citing the Hebbian principle of learning, neurons that 'fire together wire together', van der Kolk writes:

> When a circuit fires repeatedly, it can become a default setting—the response most likely to occur. If you feel safe and loved, your brain becomes specialized in exploration, play, and cooperation; if you are frightened and unwanted, it specializes in feelings of fear and abandonment (van der Kolk 2014, 56)[14]

Repeated exposure to stressful events establishes well-worn neurological pathways. In addition, prolonged stress is correlated with other factors, including elevated cortisol levels and irregularities in the hippocampus, a key structure for memory (Sapolsky 2004, 215–225).

But it is not just prolonged exposure to adverse circumstances that is potentially transformative. Cozolino (2014, 403) suggests that even a singular experience can have enduring effects. He writes:

> In a single highly charged affective moment, any of us can learn to be terrified of something for the rest of our lives. On the other hand, learning not to be afraid can take months or even years of reconditioning.

In this light, consider the potential impact of the following incident that occurred in July 2000 in Massachusetts. After a youth hockey practice, Thomas Junta, father of a participating player, beat to death coach Michael Costin in the presence of the sons of both men. Ironically, Junta was supposedly bothered by the rough play that Costin allowed during the hockey drills.[15]

Emotional trauma warrants our moral and prudential concern. 'Psychological shock' is a form of harm. The proximal cause is an underlying injury to the brain. These brain injuries affect significant interests of the individual, both in the short term and long term. These affected interests include the quality of conscious states and decision-making capacities. To the extent that the behaviors of individuals in sporting contexts contribute to emotional trauma and brain injury, they are cause for moral concern. But is there reason to think that emotional trauma is a *widespread* problem in sport? Some evidence supports that conclusion.

4. Emotional Trauma and Sport

Over the past several years, numerous books have recorded stories of athletes with disturbing emotional profiles. These works include Joan Ryan's *Little Girls in Pretty Boxes: The Making and Breaking of Elite Gymnasts and Figure Skaters* (2000), a story of psychological abuse, anorexia, bulimia, and even death, Laura Robinson's *Crossing the Line: Violence and Sexual Assault in Canada's National Sport* (1998), Jennifer Sey's *Chalked Up: Inside Elite Gymnastics' Merciless Coaching, Overzealous Parents, Eating Disorders, and Elusive Olympic Dreams* (2008), Andre Agassi's *Open: An Autobiography* (2010), and, in a different vein, former triathlete Scott Tinley's doctoral dissertation entitled *Seeing Stars: Emotional Trauma in Athlete Retirement: Contexts, Intersections, and Explorations* (2012).[16]

It is important to distinguish between cases with different types of profiles. In some instances, sport may be an arena in which antecedently acquired pathology is displayed. In other cases, participation in sport may be a major contributing factor to emotional trauma, or may exacerbate pre-existing emotional vulnerability. A further complicating factor is that it is difficult to assign contributions to nature and nurture with precision, and we must allow for individual differences in responses to environmental factors. Former gymnast Jennifer Sey (2008, xi) writes in *Chalked Up*: 'I was born with a competitive ire and near-manic ambition. Often this predisposition provides an edge in a highly competitive culture. At times, it morphs into self-destructiveness…Gymnastics was the first excuse for me to turn on myself'. Taken at face value, Sey's comment suggests that, in her case, participation in gymnastics exacerbated a pre-existing condition.

In any case, we can identify numerous factors that potentially contribute to emotional trauma in sport. These include verbal abuse, sexual abuse, and other forms of physical abuse from coaches, pressure and criticism from parents and fans, hazing rituals by teammates, perceived failure, injuries that require a hiatus or permanent departure from a sport, and retirement from a lifetime of identity shaping competition. In some cases, a confluence of contributing factors may be in play.

Though people may experience emotional trauma at any point in life, it is noteworthy that some participants in sport take part in pressurized, elite-level competition during critical stages of brain development. In their book entitled *The Teenage Brain: A Neuroscientist's Guide to Raising Adolescents and Young Adults* (2015), neuroscientist Frances E. Jensen and co-author Amy Ellis Nutt write of the until recently 'relatively neglected' teenage brain (3) and claim that this 'age window' contains 'unique vulnerabilities' (4). They add:

> Because of the flexibility and growth of the brain, adolescents have a window of opportunity with an increased capacity for remarkable accomplishments. But flexibility, growth, and exuberance are a double-edged sword because an 'open' and excitable brain can be adversely affected by stress, drugs, chemical substances, and any number of changes in the environment. And because of an adolescent's often overactive brain, those influences can result in problems dramatically more serious than they are for adults. (Jensen and Nutt 2015, 23)

According to Jensen and Nutt (2015), stress functions in a different way in adolescents than in adults 'and the effects of stress on learning and memory in teenagers can predispose them to mental health problems, including depression and post-traumatic stress disorder (PTSD)'. One significant difference is in how adults, on the one hand, and adolescents, on the other hand, respond to the hormone tetrahydropregnanolone (THP). The hormone is secreted when stress is experienced. In adults the hormone works to lower anxiety, but in adolescents it increases anxiety. The result is that 'anxiety begets even more so in teens' (Jensen and Nutt 2015, 22).

The idea that there are critical periods of brain development, with heightened opportunities and risks, suggests a further topic ripe for neuroethical reflection, with potential implications for reform in sport. But individuals might also pose caveats to the issues that I have raised.

5. Some Potential Objections and Responses

Thus far I have attempted to show that emotional pathology can be the result of traumatic interrelationships, and that emotional pathology is, at the least, accompanied by injury to the brain. Furthermore, I have provided evidence that emotional trauma is perhaps not uncommon in the context of sport. *Prima facie*, this would seem to constitute a serious issue. Some individuals, however, might still have reservations about fully embracing this claim. Aside from objections potentially deriving from metaphysical commitments, other concerns might be raised to the claims of this paper. One potential objection is that growth requires stress on a system. Sport is rough and tumble. It has built-in adversity. But these qualities allow participants to develop and exhibit the grit, determination, and resilience that will serve them well in both sport and beyond.[17] Thus, there should be no coddling in sport. There is some truth to this objection, but there is a difference, though in some cases perhaps ill-marked, between some stress to the system, on the on hand, and injury caused by unrelenting stress and abuse.[18]

Some may be undeterred by this response. Even if emotional trauma in sport is connected to brain injury, this is not unequivocally bad, some may say, since compensatory advantages may be conferred by the injury. This consequentialist-style argument would find justification for injury in an offsetting good. At times it is the case that there are remarkable trade-offs with brain injuries, as discussed in a book entitled *The Paradoxical Brain*, edited by Narinder Kapur (2011, 40–73).

Kapur outlines a number of 'paradoxical cognitive phenomena' linked to neurological conditions (Kapur, 2011, 40; see pp. 40–64 for elaboration). Kapur writes:

> In neurological conditions, the major sets of paradoxical cognitive phenomena generally take one of five forms: (1) enhanced cognitive performance of neurological patients vis-à-vis neurologically intact individuals ('lesion facilitation'), and (2) alleviation or restoration to normal of a particular cognitive deficit following the occurrence of a second brain lesion ('double-hit recovery'). (3) A third set of paradoxical cognitive phenomena represents what may be termed 'hinder-help effects', where a variable that produces facilitation or detriment of performance in healthy participants results in opposite effects in neurological patients. (4) A fourth form of paradox relates to anomalies in the usual relationship between the presence/size of a brain lesion and the degree of cognitive deficit ('lesion-load paradox'). (5) A fifth paradox is where there may appear to be direct or indirect benefits for long-term neurological outcome as the result of specific cognitive deficits being present ('paradoxical positive outcome'). (Kapur 2011, 40)

Thus in a variety of ways there is evidence of what has been called 'post-traumatic growth' (Kapur, 63). But while 'post-traumatic growth' is possible, the consequentialist argument in which it figures prominently is fraught with difficulty when it comes to emotional trauma in sport. This is not least the case because individuals may react differently to the same stimuli. There is no *guarantee* that an off-setting good will result from emotional trauma, or that an athlete would or could in an informed way agree to the trade-off beforehand. These could be, after all, 'transformative experiences' in Paul's (2014) sense of the term, which to some extent take us to unknown territory.

Paul (2014) argues that decisions to undergo transformative experiences of a certain type pose challenges to our abilities to make rational choices. The challenge is twofold.[19] To illustrate the difficulties Paul explores the choice as to whether to become a vampire or not. The decision is hindered in the first place because it is epistemologically opaque. Since one does not know what it is like to be a vampire, one is not in the position to compare the 'subjective values' (Paul 2014, *passim*) attached to remaining as one is, on the one hand, and becoming a vampire, on the other.[20] There is an additional confounding factor. Once one becomes a vampire one's preference structure may change. Once one's preference structure changes, one may prefer being a vampire to being a non-vampire; nevertheless, from one's prior pre-vampire perspective, one would have perhaps preferred to remain a non-vampire.[21] Is it possible to make a rational choice in such cases? Paul suggests that the solution may be to go to a higher level. That is, one makes the decision not on the basis of a comparison of first-order subjective values, but rather on the basis of valuing the discovery or revelation of what it is like to be a vampire.[22]

It seems that at least some brain injuries are transformative in the sense outlined by Paul. But if brain injuries connected to emotional trauma are transformative in this way, a rational choice to undergo them would mean choosing them because one values learning what it is like to have a brain injury. But this is a problematic framework for rational decision-making, particularly since severe brain injuries not only may compromise one's autonomy, but also one's ability to fully 'appreciate' the revelatory condition of having a brain injury. In some cases, the transformations could be radical and irreversible.[23] Malabou (2012) writes: 'Changes caused by brain lesions… frequently manifest themselves as an *unprecedented metamorphosis* of the patient's identity'. In limit cases, one may even be unconscious or dead as a result of the injury.

Thus, it may be shortsighted to say that athletes autonomously subject themselves to brain injuries in sport.[24] *A fortiori*, an appeal to autonomy fails to account for cases involving children and teenagers who are vulnerable and who are unable to give fully informed consent. In some cases they may be pressured into participation. It is worth noting in this context that the word 'injury' derives from the Latin 'injuria', which referred to injustices and violations of rights (Feinberg 1987, 107).

Another potential rejoinder is to say that a 'bracketed morality' applies to sport. Sport is not the real world, and so everyday morality does not hold here (Shields 2010). But the considerations raised in this paper reveal the significant limitations of this approach. Whatever circumscribed application the argument may have, participation in sport has life-altering consequences, some of which are irrevocable. Given this, it seems fitting that the burden of proof should fall on those who argue for bracketed morality, particularly if the stakes are high.

So, what about sports like football, soccer, and boxing? While brain injuries that occur in these sports may be mostly unintended, they are nevertheless foreseen. Yet we don't prohibit these sports or radically overhaul them. What reason is there to think that there is sufficient concern to generate significant change in our approach to emotionally charged brain injury in sport? A twofold response is warranted. First, the capitulation reflected in this objection obscures the is-ought distinction. When it comes to our moral obligations, we do not discover them by merely consulting present opinion or registering the current motivational climate. Second, the pessimistic tone of the objection may be unwarranted. That is, it may underestimate the amount of change that is on the horizon. Take the case of North American-style football. Football is undergoing reform, and the conversation about needed reforms has just started. Even tough-minded Mike Ditka, former National Football League player and former coach of the Chicago Bears joined the chorus of those who have said that they would not want their children to play football (Davidson 2015). As Bob Dylan might say, 'the times they are a-changin'. Thus it behooves us to consider brain injury across the board at this propitious moment.

6. Conclusion

In this paper, I have argued that severe emotional disturbance that occurs in the aftermath of emotional trauma experienced in the context of sport is evidence of underlying brain injury. While attention is currently, and rightly so, focused on concussions that occur in sport, we should not neglect brain injuries that occur by means other than physical blows to the head or otherwise jarring the brain. The brain pathologies that accompany emotional disturbance may be difficult to characterize or locate. Nevertheless, manifest psychological shock is a clue to their existence. There is evidence that suggests that emotional trauma has been experienced in the context of sport on a widespread basis. But regardless of how pervasive emotional trauma in sport may be, its presence warrants attention.

Participation in sport can be a transformative experience. But as Paul (2014) notes, with respect to certain kinds of personal transformations, we cannot know what the transformation will be like prior to being transformed. We do have some idea of the kinds of transformations that occur through changing one's brain through concussions or emotionally traumatic experiences that have other origins. I have argued that emotionally traumatic experiences can result in brain injury as well as psychological harm. The individuals who have given voice to their own traumatic experiences merit our compassion and moral concern, as do the affected, but voiceless ones whose stories we have yet to hear.

The intricate workings of the brain are still wondrous and confounding, but it is no longer feasible to dismiss the brain as a 'black box' that we can set aside in our discussions. Instead, it is incumbent on us to make present the absent brain, along with its varieties of vulnerabilities.

Notes

1. Accessible and informative accounts of some of the issues include Cantu (2013) and Carroll and Rosner (2012).
2. Mild TBI symptoms, TRAUMATIC.BRAIN.INJURY.COM, www.traumaticbraininjury.com/symptoms-of-tbi/mild-tbi-symptoms/, accessed 21 July 2016, states: 'A traumatic brain injury (TBI) can be classified as mild if loss of consciousness and/or confusion and disorientation is shorter than 30 minutes. While the MRI and CAT scans are often normal, the individual has cognitive problems such as headache, difficulty thinking, memory problems, attention deficits, mood swings and frustration. These injuries are commonly overlooked. Even though this type of TBI is called "mild", the effect on the family and the injured person can be devastating.'
3. When is emotional disturbance sufficiently pronounced to warrant positing an underlying brain injury? It is difficult to provide a fine-grained analysis here. But the same could be said with respect to the ordinary discourse sense of injury. While there are borderline cases in both instances, other cases are clearer. Note that while my focus is on the brain, emotional trauma also affects other parts of the body, such as the heart. I am grateful to Jennifer Hardes for pointing this out.
4. See LeDoux (2015, 203–232), for a fuller account of how 'emotional consciousness' arises.
5. My arguments are meant especially for those who espouse forms of dualism that fail to recognize that our mental states are grounded in and caused by brain states and for those in general who, to various degrees, are tempted to discount emotional problems as 'merely' psychological states.
6. Note that while property dualism is a weaker form of dualism than substance dualism, property dualism also raises objections. These include charges of causal overdetermination and epiphenomenalism.
7. I am grateful to Nicole Vincent, who posed questions to me regarding what really matters in this discussion. In particular, is it not the case that the psychological and behavioral manifestations are what fundamentally matter to us? Is it not the case that the harm that we are really concerned about is at the psychological level? She posed the following functionalist-style thought experiment to support this view. Suppose we could substitute electronic devices for our brains and yet maintain the same psychological profiles. If so, this would show that brains are not crucial to the discussion. We are concerned about brains because we think that they stand in a causal relationship to our psychological and behavioral traits. My response is threefold. First, if some form of token identity physicalism is the accurate view, then while it might not be the case that brains per se are critical, my brain will be critical in my present physical instantiation. A damaged brain will result in a damaged psyche. Even if the brain is only causally related to my psychological states, then they are closely connected to *my* psychological condition. Second, interventions on behalf of my welfare may need to be bottom up, as, for example, in the case of a tumor that does not respond to talk therapy. A third point is related to psychology, and not just metaphysics. If one sees the intimate connection between our psychological states and approximately three-pound squishy mass of matter inside our heads, one may become more aware of the fragility of the good life, and of human existence

in general. I am indebted to William FitzPatrick for a comment that was helpful for my framing of the third point.
8. In this regard, consider neuroscientist Sebastian Seung's (2013, xxi) reference to 'connectopathies, abnormal patterns of neurological activity'.
9. Cozolino (2014, 383) dates the dawning realization that experience affects brain structure and function to the 1990s.
10. I am indebted to Cara Wellman for drawing this issue to my attention.
11. See Doidge (2007, 2015).
12. Even in cases of transitory psychological shock it is plausible to say that brain injury of varying degrees of severity and varying durations is sustained. I am indebted to Mike McNamee for calling my attention to the need to address this point.
13. LeDoux (2003, 98) attributes the origin of the term 'engram' to the German scientist Richard Semon in 1904.
14. Agassi's (2010, 38) comments on the aftermath of a lost tennis match as a child suggest that already by age eight well-worn pathways were present in his brain. He writes:

    After years of hearing my father rant at my flaws, one loss has caused me to take on his rant. I've internalized my father-his impatience, his perfectionism, his rage—until his voice doesn't just feel like my own, it is my own. I no longer need my father to torture me. From this day on, I can do it all by myself.

    I make this point using this text also in Fry (2013, 159).
15. See '"Hockey dad" gets 6 to 10 years for fatal beating', CNN.com/LAWCENTER, January 25, 2002. https://www.cnn.com/2002/LAW/01/25/hockey.death.verdict/index.html?_s=PM:LAW Accessed 3 July 2016.
16. Other relevant discussions include Nack and Yaeger (1999) on coaches and sexual abuse and Wolff, with Shute (2015) on abusive coaching.
17. On the importance of grit see Duckworth (2016). Research by sports psychologist Tim Woodman and others found a correlation between being an elite athlete and having had a significantly negative experience earlier in life. In addition, many elite athletes also experienced another critical episode (positive or negative) during their careers. See the article by Nuwar (2016).
18. I am indebted to Anna Kalinovsky for pointing out this distinction. Note that it is beyond the scope of this paper to establish criteria for adjudicating hard cases. But one does not have to be able to produce such criteria in order to claim that some morally problematic cases of emotional abuse are not hard to recognize. One basis for claiming that such cases are morally problematic is Kant's (1964, 101) injunction that one 'should treat himself and all others *never merely as a means*, but always *at the same time as an end in himself*. It is also beyond the purview of this paper to establish policies. The limited intent of this paper is to show that a problem exists and that it is not 'merely' psychological in nature.
19. See in particular Paul (2014), chapter 2, 5–51.
20. The reference to what it is like recalls the classic piece by Nagel (2004), 'What is it Like to be a Bat?'. Nagel's response is that we have no idea of what it is like to be a bat. Nevertheless, Nagel thinks that there is something that it is like to be a bat, and that there being 'something that it is like' is the hallmark of consciousness.
21. Thus Paul (2014, 17) says that the kind of case she has in mind can be both 'epistemically transformative, giving you new information in virtue of your experience,' and 'personally transformative, changing how you experience being who you are'.
22. On the higher order approach see especially Paul (2014), chapter 4, 105–123.
23. Perhaps not all personal transformations are profound. Some may propose that if the change is superficial, it is a trivial matter. Thus it is important to show that in some instances emotional trauma is profoundly transformative. This helps establish the moral significance of these cases. A comment by an anonymous reviewer prompted me to offer this gloss.
24. For a thought-provoking discussion of athletes subjecting themselves to risk and the complexities of autonomy see Lurie (2006).

Acknowledgments

Versions of this paper were presented at the 43rd Annual Conference of the International Association for the Philosophy of Sport, Cardiff, Wales, UK September 2015, and the Ethics and the Brain Conference, Flint, Michigan, May 2016. I am grateful for feedback I received from participants at these conferences. I am also grateful for feedback that I received at a roundtable discussion at the Poynter Center for the Study of Ethics and American Institutions at Indiana University and to comments from Elizabeth N. Agnew, Jennifer Hardes, Mike McNamee, Andrew Edgar, and two anonymous reviewers for *Sport, Ethics and Philosophy*.

Disclosure Statement

No potential conflict of interest was reported by the author.

References

AGASSI, A. 2010. *Open: An autobiography*. New York, NY: Vintage Books/Random House. First published 2009 by Alfred A. Knopf.

CANTU, R. 2013. *Concussion and our kids: America's leading expert on how to protect young athletes and keep sports safe*. Boston, MA: Mariner Books/Houghton Mifflin Harcourt.

CARROLL, L., and D. ROSNER. 2012. *The concussion crisis: Anatomy of a silent epidemic*. New York, NY: Simon & Schuster.

CORKIN, S. 2013. *Permanent present tense: The unforgettable life of the amnesic patient H.M.* New York, NY: Basic Books/Perseus Books.

COZOLINO, L. 2014. *The neuroscience of human relationships: Attachment and the developing social brain*. 2nd ed. New York, NY: W.W. Norton & Company.

DAVIDSON, K. 2015. Even Mike Ditka thinks football is too dangerous, *Chicago Tribune*, 21 January. Available at http://www.chicagotribune.com/news/opinion/commentary/chi-mike-ditka-football-dangerous-son-20150121-story.html (accessed 11 July 2016).

DESCARTES, R. 1993. *Meditations on first philosophy in which the existence of god and the distinction of he soul from the body are demonstrated*. 3rd ed., translated by Donald A. Cress. Indianapolis, IN: Hackett Publishing Company. First published in 1641.

DOIDGE, N. 2007. *The brain that changes itself: Stories of personal triumph from the frontiers of brain science*. New York, NY: Penguin Books.

DOIDGE, N. 2015. *The brain's way of healing: Remarkable discoveries and recoveries from the frontiers of neuroplasticity*. New York, NY: Viking/Penguin Group.

DUCKWORTH, A. 2016. *Grit: The power of passion and perseverance*. New York, NY: Scribner/Simon & Schuster.

FAINARU-WADA, M., and S. FAINARU. 2013. *League of denial: The NFL, concussions, and the battle for the truth*. New York, NY: Crown Archetype/Crown Publishing Group.

FEINBERG, J. 1987. *The moral limits of the criminal law volume 1: Harm to others*. New York, NY: Oxford University Press. First published 1984 by Oxford University Press.

FINGER, S. 1994. *Origins of neuroscience: A history of explorations into brain function*. New York, NY: Oxford University Press.

FRY, J.P. 2013. The neuroethics of coaching. In *The ethics of coaching sports: Moral, social, and legal issues*, edited by Robert L. Simon. Boulder, CO: Westview Press/Perseus Books Group: 151–166.

GLANNON, W. 2013. *Brain, body, and mind: Neuroethics with a human face*. Oxford: Oxford University Press. First published 2011 by Oxford University Press.

HEIL, J. 2013. *Philosophy of mind: A contemporary introduction*. 3rd ed. New York, NY: Routledge/Taylor & Francis.

'HOCKEY DAD' GETS 6 to 10 years for fatal beating. *CNN.com/LAWCENTER*, 25 January 2002. http://www.cnn.com/2002/LAW/01/25/hockey.death.verdict/index.html?_s=PM:LAW (accessed 3 July 2016).

JENSEN, F.E. and A.E. NUTT. 2015. *The teenage brain: A neuroscientist's survival guide to raising adolescents and young adults*. New York, NY: Harper/Harper Collins.

JOHNSTON, E., and L. OLSON. 2015. *The feeling brain: The biology and psychology of emotions.* New York, NY: W.W. Norton & Company.

KANT, I. 1964. *Groundwork of the metaphysic of morals.* Translated by H.J. Paton. New York, NY: Harper Torchbooks/Harper & Row.

KAPUR, N. 2011. Paradoxical functional facilitation and recovery in neurological and psychiatric conditions. In *The paradoxical brain*, edited by Narinder Kapur, with Alvaro Pasual-Leone, Vilayanur Ramachandran, Jonathan Cole, Sergio Della Sala, Tom Manly, and Andrews Mayes. Cambridge: Cambridge University Press: 40–73.

KIM, J. 2011. *Philosophy of mind.* 3rd ed. Boulder, CO: Westview Press/Perseus Books Group.

LEDER, D. 1990. *The absent body.* Chicago, IL: The University of Chicago Press.

LEDOUX, J. 2015. *Anxious: Using the brain to understand and treat fear and anxiety.* New York, NY: Viking/Penguin Random House LLC.

LEDOUX, J. 1998. *The emotional brain: The mysterious underpinnings of emotional life.* New York, NY: Touchstone/Simon & Schuster.

LEDOUX, J. 2003. *Synaptic self: How our brains become who we are.* New York, NY: Penguin Books/Penguin Group. First published 2002 by Viking Penguin.

LURIE, Y. 2006. The ontology of sports injuries and professional medical ethics. In *Pain and injury in sport: Social and ethical analysis*, edited by Sigmund Loland, Berit Skirstad, and Ivan Waddington. London: Routledge/Taylor and Francis Group: 200–210.

MALABOU, C. 2012. *The new wounded: From neurosis to brain damage.* Translated by Steven Miller. New York, NY: Fordham University Press. First published 2007 as *Les nouveaux blesses* by Bayard editions.

MILD TBI SYMPTOMS, *TRAUMATIC.BRAIN.INJURY.COM*, WWW.TRAUMATICBRAININJURY.COM/SYMPTOMS-OF-TBI/MILD-TBI-SYMPTOMS/ (ACCESSED 21 July 2016).

NACK, W. and D. YAEGER. 1999. Every parent's nightmare. *Sports illustrated* 13 September, 91 (10): 40–53.

NAGEL, T. 2004. What is it like to be a bat? In *Philosophy of mind: A guide and anthology*, edited by John Heil. Oxford: Oxford University Press: 528–538. First published in *Philosophical Review* 83 (1974).

NUWAR, R. 2016. The right stuff: What psychological and physical traits separate the world's best athletes from the rest of us? *Scientific American Mind: Behavior. Brain Science. Insights* July/August: 38–44.

PAUL, L.A. 2014. *Transformative experience.* Oxford: Oxford University Press.

ROBINSON, L. 1998. *Crossing the line: Violence and sexual assault in Canada's national sport.* Toronto: McClelland and Stewart.

RYAN, J. 2000. *Little girls in pretty boxes: The making and breaking of elite gymnasts and figure skaters.* New York, NY: Warner Books/Time Warner. First published 1995.

SAPOLSKY, R.M. 2004. *Why zebras don't get ulcers.* 3rd ed. New York, NY: St. Martin's Griffin.

SEUNG, S. 2013. *Connectome: How the brain's wiring makes us who we are.* Boston, MA: Mariner Books/Houghton Mifflin Harcourt.

SEY, J. 2008. *Chalked up: Inside elite gymnastics' merciless coaching, overzealous parents, eating disorders, and elusive Olympic dreams.* New York, NY: William Morrow/HarperCollins.

SHIELDS, D. 2010. Rethinking competition. TrueCompetition.Org. Reclaiming competition for excellence, ethics, and enjoyment. 14 April. Available at Truecompetition.org/resources/rethinking-competition/ (accessed 26 July 2016).

TINLEY, S. 2012. Seeing stars: Emotional trauma in athlete retirement: Contexts, intersections, and explorations. Doctoral diss., Claremont Graduate University.

VAN DER KOLK, B.A. 2014. *The body keeps the score: Brain, mind, and the body in the healing of trauma.* New York, NY: Viking/Penguin Group.

WOLF, A. and L. SHUTE. 2015. Abuse of power. *Sports Illustrated*, 28 September: 50–55.

On the Compatibility of Brain Enhancement and the Internal Values of Sport

Alberto Carrio Sampedro ⓘ and José Luis Pérez Triviño

ABSTRACT

Elite athletes are characterized by their high level of performance in sport. Since the very beginnings of sport, it has been understood that physical and physiological abilities influence the performance of athletes. Advances in scientific knowledge, especially sport psychology and neuroscience, seem to confirm this intuition and consequently it is possible to characterize elite athletes as having an extraordinary combination of physical and mental abilities. Techniques and substances that contribute to enhancing physical characteristics of athletes have also been well known for ages. But it is now possible to make use of other techniques and substances that not only enhance physical abilities but also cognitive capabilities, which seem to require greater consideration given their direct impact on the athlete's brain. In this article, we examine two such techniques, cognitive enhancers and transcranial stimulators, and highlight the potential advantages and drawbacks that applying each one may have on sport. Given the relative novelty of these enhancement techniques and substances and the absence of conclusive evidence regarding their short- and long-term effects, we deem that their use ought to be strictly governed by cautionary principles. But due to that same lack of evidence, we believe that the possibility of examining the feasibility of applying these techniques to sport should not be denied.

1. Introduction

According to Hoberman, performance in sport can be defined as all the psychological or physiological efforts that can be quantified or assessed in physical or psychological terms (Hoberman 64). Without a doubt, the effort put in by elite athletes and their level of control in strenuous situations exemplify the epitome of sport performance. This almost inhuman psychological and physiological performance ability embodies, to use McNamee's ideas, the genuine ideal of the athlete.[1] It can be said, therefore, that the performance of elite athletes is the result of an extraordinary confluence of physical and psychological abilities. We admire athletes not only for their physical strength, their extraordinary ability to push through their limits, their self-confidence and their ability to adapt to unpredictable situations, but also for their *psychological* orientation towards success. All of these abilities are defining characteristics of what has been call the 'athletic identity' (AI).[2]

In this article, we assume that cognitive abilities play a decisive role in sport performance, and that, as a result, it is important to analyse the impact and viability of new cognitive enhancement techniques such as neuroenhancers and transcranial stimulators as applied to athlete performance. Thus, to provide a roadmap for this article, in Section 2, we analyse the relationship among athletic identity (AI), training and technological enhancements applied to sport. Next, in Section 3, we provide a general overview of cognitive enhancement in sport. In Section 4, we deal with neuroenhancers and in (5) we turn to transcranial stimulators. In Section 6, we analyse some of the objections and normative problems related to the use of these techniques in sport. We will then propose an interpretation that makes the restricted use of these techniques compatible with the internal values of sport.

2. The athletic personality and technological, physical and psychological enhancements

Physical abilities are a decisive factor in sport. When thinking of athletes training, it is likely that the first image that comes to mind is, as we have already mentioned, the great physical effort made when they follow strict training regimes. Certainly athletes, and especially elite athletes, go above and beyond the limits of healthy physical exercise as recommended by the World Health Organization (WHO).[3] Consider, for example, that in accordance with the 'Nomenclature for performance achievement levels' (NPAL), athletes who are Olympic gold medal winners or world champions are classified as 'super elite' while those who finish in second place are characterized as 'elite'.[4] There is little doubt that any of the athletes who fit into these categories possess quite a rare combination of physical and psychological traits. One of the first things that all athletes learn at the very beginning of their careers is that winning requires hard work and the demonstrated ability to suffer through and adapt to inherently adverse circumstances in all competitive situations. There are certainly many restrictions that an athlete's life is subject to: rigorous training sessions, strict diets, injuries and concentration on the desired goal. The attitude and action of overcoming these obstacles is called 'resilience'.[5] In the sport world, 'resilience' can be defined as a dynamic psychosocial process which at the same time (i) implies interacting with others and reinforcing the values of mutual support and friendship, particularly in team sports and (ii) is temporary and factually dependent, given that it varies depending on the competition schedule and factual circumstances, such as injuries.[6] It is obvious that the lifestyle of any of those athletes is not exactly enviable.[7] This is even more obvious given the fact that if the World Anti-Doping Agency requirements for maintaining a 'clean' biological passport are taken into account, athletes are in fact pressured to place the demands of their profession over taking care of their own health.

To meet these demands, athletes tend to resort to applying technology to their training, a trend that has been growing in recent years. The elaborate design of modern athletic apparel and shoes, the machines that allow athletes to build up strength and endurance, the wind tunnels used to rehearse body techniques to reduce air resistance, and many other innovations are simply the most well-known examples.

But physical performance cannot be equated with raw performance. To produce best results, physical performance must be accompanied by strict concentration and training for competition. Said another way, the effort needed to take full advantage of an athlete's maximum potential must be rationalized. Thus, this is where AI plays an essential role.[8] Let us examine this concept in greater detail.

Athletes can be defined by the degree with which they identify with their activity.[9] In this way, those who possess a strong Athletic Identity are more willing to submit themselves to the restrictions required by the practice of the sport. That is to say those who possess this strong athletic personality are better psychologically prepared to deal with the rigours of competition. As we will see further on, this characteristic is essential to success in highly competitive sport.[10]

Thus, if AI lays the groundwork for the psychological motivation of the athlete, the ends seems to be as important as the means employed to attain it. This raises the question of whether the ends and means are completely incompatible, or if, on the contrary, to what extent they are compatible.

The incompatibility thesis can be defended from different perspectives. But once the importance of the ends is conceded—to increase the level of motivation of the athlete, scientific and technological advances to improve athlete motivation offer great opportunity. As we have just seen, this is what occurs with preparation and physical performance thanks to the scientific and technological advances that enhance muscle endurance or increase aerodynamics. At the end of the day, all these means are human artefacts that fall into the category of technology in sport.[11] Thus, since 'technology is a necessary condition for many sports to arise at all'[12]; the question is which technology should be permissible to use in sport to attain the very goals of sport?

Another perspective of incompatibility is the one that highlights that cognitive enhancement techniques put the continuity of internal values of sport at stake and they consequently distort the nature of the practice.[13] The question that follows is why we must be more averse to the scientific advances applied to enhancing athletes' brains (and their cognitive functions), than to other kinds of enhancement. Such is the case of physical enhancement with all the nuances that have emerged in discussions of doping.[14]

The compatibility thesis is not blind to the implications of cognitive enhancers in sport. It only points out that there is little doubt that once the physiology of the brain is known and the areas that influence motivation and other cognitive abilities that affect athlete performance have been mapped,[15] it is possible to stimulate them to enhance that performance. The issue to which we alluded earlier, is which of these theses offers a more promising perspective on sport and, in any case, to what extent these means are compatible with the ends and goals of sport.

The repercussions that psychological abilities have on sport performance have been well known for quite some time (Hoberman). Among these abilities, it is worthwhile to differentiate between moods—emotional abilities in the broad sense of the term and cognitive abilities. One practical example with regard to the former is the possibility of treating certain emotional states, like depression, extreme shyness or phobias, in order to better the quality of life of those who suffer from these states. It seems obvious that applying these techniques to the sport arena contributes to a better quality of life and improved performance of athletes. This use of psychological techniques would be included in what Loland has called technology which can 'prevent injury and protect against harm'.[16]

Yet another category would better integrate cognitive enhancements in the strict sense of the term, i.e. those which aim to go beyond dealing with the above-mentioned treatment cases and directly affect performance in sport. In other words, according to Loland, they constitute 'performance-enhancing technology in sport'.[17] The following table provides an overview of the different types of intervention and their possible application to sport.

Abilities:	Therapy	Enhancement
Physical.	1	2
Cognitive.	3	4

As we mentioned above, this article focuses on the fourth case, i.e. on the enhancement of cognitive abilities given that advances in the field of sport psychology have opened up new prospects for their treatment and enhancement. It is for this reason, as we also alluded to above, that we will deal with the impact of two of the techniques currently used in sport psychology on sport enhancement, the so-called cognitive enhancers and cranial stimulators.

But before going on to scrutinize the challenges that these techniques present for sport, perhaps it would be better to make a few basic assumptions explicit.

(1) The physical and psychological abilities of athletes, despite being sine qua non for elite athletes, are not always the determining factor of success. There are cases in which luck plays the key role.[18]
(2) It is necessary to consider that genetic and social factors are involved in the possession and development of these abilities.
(3) The impact of the natural lottery, social status and luck is uneven, depending on the type of competition that is being dealt with.[19]

The analysis to be carried out focuses on the growing interest that the open debate in psychology and neuroscience has sparked in the sport arena with regard to the mind–body duality,[20] but given that it is not possible to deal with such a broad scope of issues, we will restrict our focus to technical and scientific innovations which are liable to be applied to enhancement of sport performance.

3. Cognitive abilities in sport

The zeal to improve performance in sport has been a constant throughout the history of sport. And given that sport performance, as we have just seen, is influenced by both physical and cognitive factors, it is not rash to claim that the interest in improving both has been equal. But it has not been until modern times that sport has felt the reverberations of the incursion of sport psychology as a powerful instrument to improve the performance of athletes:

> The first physiologist discovered that scientific studies must operate at this border where physiology and psychology overlap. (Hoberman 1992, 157)

In the evolution of sport psychology, we see at least two stages. In the first, the fundamental concern was to balance the athlete's emotional states in such a way that performance would be improved. Hoberman confirms that 'the idea of manipulating the mind—a popular concern in our times'—occurs at the beginning of the twentieth century. (Hoberman 1992, 225). The research from abnormal psychology in this area focused on better management of stress, burnout and end-of-career anxiety (Tamorri 2004, 5).

The second stage can be identified by an overall expansion of sport psychology that allowed a glimpse at the potential to optimize mental training and mood, which are highly influential in sport performance.

There are at least two relevant reasons to deal with the impact of cognitive improvement on sport performance and, consequently, the structure of competition. This is because in certain disciplines, the cognitive aspect has special relevance. An extreme example of this is chess, but other sports are not far behind: these others require a high degree of interdependence between competitors to carry out complex strategies, as is the case in most of team sports.

The second reason is more significant. It concerns the link between physical and cognitive abilities on sport performance (Foddy 2011). Even if it is true that physical abilities are those that can be quantified in terms of sport performance, it is not less true, as Bennet Foddy highlights that

> all such variations (physical actions) are mediated, at least in part, by the actor's brain, spinal cord, and peripheral nervous system. Neurological systems play a role in determining how far we throw a javelin, how deeply we breathe while swimming … and how long we can withstand the pain of endurance cycling.

Cognition is understood as the process employed by an organism to organize information, a process which includes the following abilities: (i) acquisition (perception); (ii) selection (attention); (iii) interpretation (understanding) and (iv) retention (memory). In accordance with this, cognitive enhancement can be defined as any increase in information-processing abilities involved in the cognition process by any means or system, internal or external (Sandberg 2011, 71). Given this characterization, it is not difficult to highlight the intimate connection between these abilities and the functions that sportspersons carry out on a daily basis in their athletic workouts. It is, therefore, unsurprising that the interest in cognitive enhancement has done nothing but increase in the sport arena. There are even popular sayings which imply that these abilities are the reason behind winning:

> The difference between winning and losing is 99% psychological.

> 90% of sports is mental and the other half is in the head.

Although these claims may be exaggerated, it is clear that there have been different psychological methods and techniques developed in recent decades that improve sport performance.[21] The advances in sport psychology (LeUnes, 2011, 201), cognitive sciences and neuroscience thus seem to have elements that should be taken into account.

Just consider the spectacular development that has been seen in neuroscience and the importance that its progress has had in promoting more accurate knowledge of brain function. The tools that it offers allow the intuitions of sports psychologists to be explained with a little more clarity, in this way linking motor functioning directly to the brain.[22]

There are even those who argue that advances in neuroscience entail revolutions similar to those which were sparked by Galileo in physics or Darwin in biology. Taking all that into account, sport cannot stand by the wayside in the face of those advances. But rather the opposite is true: it should work together with them as they allow exploring unimaginable possibilities (Tamorri 2004, 10).

Perhaps the main contribution of neuroscience to sport will be the possibility of establishing a greater degree of accuracy of the relationship between brain and motor functions and their eventual enhancement.

This is what is occurring with brain stimulation, whether it be visual, auditory or of the video variety, to locate areas that process information. Once they have been located and their complex functioning is understood, there is nothing to stand in the way of improving it, to the best of our abilities. Tamorri, for example, argues that the structure of a champion can be suitably identified:

> A champion is a blend of muscular reaction and biomechanics, developed through a delicate, fine and complex process of information recovery, decoding and programming that is found in his brain, in his biology, in his neurotransmitters and finally, in his cognitive processes, organic capabilities but also the emotional, cultural and practical ones are the reasons behind one response or another. (Tamorri 2004, 9)

According to this author, neuroscience lends to sport:

> The knowledge of the molecular and neurochemical mechanisms at the base of motor memory and tactical memory, the athlete's ability to adapt to diverse situations by quickly resynchronizing biological rhythms after jet lag or the ability to take advantage of situations such as the release of emotional states like happiness, pain, frustration, enthusiasm, disappointment, or even the plasticity in the process that allows the nervous system after it finishes to form what are likely new synapses that are located throughout a great number of in distinct associative areas at that foundation of learning processes. All of this would justify in any case the meaning of training. (Tamorri 2004, 10)

In addition to the detailed understanding of the complex interaction between biological and emotional processes in the brain, N. Davis has recently suggested that the advances in neuroscience would mean:

> that the skills and abilities underpinning sports performance can be enhanced using technologies that change the activity of the brain. These factors may include motor learning, enhanced muscular strength or reduced fatigue, or even changes to mental state or concentration. (Davis, 649)

Up to now, it seems evident that the knowledge offered by neuroscience on factors intervening in cognitive processes is extremely valuable and can be applied in the sport arena.

This being the case, the question that follows is whether this practice is justified.

We will not rush to give a hurried answer to this question, but rather in what follows we will examine some cognitive enhancement techniques and their application to sport performance.

4. Cognitive enhancers

Related to the advances which have just been mentioned, pharmaceutical laboratories have recently developed a variety of substances which can improve cognitive abilities, the so-called cognitive enhancers. These cognitive enhancers were initially designed to treat neurodegenerative diseases that arise in the aging process. But these products are also effective at improving cognitive abilities of healthy individuals. The advantages that these cognitive enhances offer are, among others, increased wakefulness and the ability to maintain high levels of attention and concentration under stressful mental conditions as well as the improvement of memory (Eronia 2012, 7). But there are also those who question their effectiveness:

> Many authors who are interested in direct brain intervention are happy to confidently assert that methylphenidate and modafinil are effective cognitive enhancers. Even when authors do not explicitly state that stimulants improve cognition, they frequently appear to assume that they

> do. However, the evidence that either drug might provide any useful form of cognitive enhancement is scant […] Even the positive finding about improvement of memory is a little difficult to translate into a real world scenario. The type of statistical analysis used was chosen because it allowed markedly different studies to be drawn together, however, the various methods used to assess memory differed markedly from study to study. (Dubljević and Ryan 2015, 26–27)

It is likely that the most well-known cognitive enhancing substances are methylphenidate (Ritalin[23]) and modafinil (Provigil); among their various effects are stronger memory and concentration abilities. Thus, they have been classified as cognitive enhancing substances (Dubljević and Ryan, 2015, 25).

The former blocks the re-uptake of dopamine, a neurotransmitter in the synapses. It can also increase the release of dopamine and noradrenalin (norepinephrine). While the connection between methylphenidate and cognitive enhancement may take place in a variety of ways, it is currently not known how the actual mechanism used by this drug works. Methylphenidate is associated with a series of adverse side effects including nervousness, drowsiness and insomnia as well as being contraindicated during pregnancy (Dubljević and Ryan, 2015, 26).

In contrast, modafinil involves short-term risks; however, its relatively recent appearance on the market prevents the long-term effects from being evaluated at this time. Despite being a weak dopamine reuptake inhibitor, modafinil's concentration after being taken orally is high enough to substantially affect dopamine reuptake, which could explain the rare in-stances of psychosis and mania related to its use.

The first trials with these substances were carried out with airline pilots and soldiers given that these drugs allowed them to improve their concentration and withstand fatigue. But with time their use has spread. Currently, it is calculated that between 5 and 15% of students in the United States have taken one of these substances with the goal of improving their academic performance. In any case, it is not just students who have tried these substances but also executives at many companies have used them with a view to mitigating fatigue, alleviating concentration deficits and avoiding burnout. Little time was needed to prove the impact that these drugs can have on special subjects, particularly those that suffer from ADHD.[24]

It was unavoidable that these substances would also impact sport. Inasmuch as they act on neurotransmitters, they improve the transmission of information and optimize physiological performance. For certain sports, these abilities are quite important and can even decisively increase sport performance; consider, for example, the effects of improvement in attention for a javelin thrower, a golfer or an archer. Nonetheless, there is insufficient research to support the claim that selective androgen receptor modulators, antiestrogens, are in fact performance enhancing. In other words, there is little data to back up their effectiveness for such a purpose.[25] In any case, there are legitimate uses of these kinds of drugs. For example, although Major League Baseball banned amphetamines in 2006, there has been a dramatic rise in the number of therapeutic use exemptions issued to players for attention-deficit disorder diagnoses, for which drugs like Ritalin and Adderall can be legitimately prescribed. In 2006, 28 players applied for the exemption, while a year later there were 103. There is growing suspicion that many of these ADD diagnoses are just excuses to get the pills.[26]

Still, scientific warnings about possible undesirable side effects, such as dependence, cannot be ignored. In all likelihood, it is for this reason that they have been included on WADA's banned listed, just as other stimulants are[27]; this is also the case for amphetamines and cocaine and was the case for caffeine in the past. At the end of the day, in all of these cases the availability of neurotransmitters in the brain is increased, inciting them to function more quickly.[28]

But in addition to these substances, there are also certain neurotechniques which are capable of improving brain functioning. There are three main types of physiological interventions on the brain (Merkel et al. 2007, 119): genetic, electromagnetic and surgical. This last one can be further differentiated into its different techniques: (a) implants or neuroprosthetics, including computer (bionic) interfaces, (b) intracranial insertion or implantation of cells to repair tissues, or cells that administer certain bioactive compounds to certain areas and (c) intracranial gene transfer techniques for heightening or diminishing the action of healing proteins.

While all of these techniques raise important issues to sport philosophy and ethics, given the limited scope/length of this article, in the following section, we will focus exclusively on electromagnetic techniques on the brain.

5. Brain stimulation techniques

In spite of the initial scepticism with which brain stimulation techniques were initially met, they have gained rather widespread acceptance in the field of sports (Goodall 2012, 7), and recent advances in neuroscience suggest that the abilities and capacities which underpin sport performance can be enhanced thanks to the use of technology that modifies brain activity. These factors may include motor learning, improved muscle strength, reduction in fatigue or even changes in mental state or concentration. As Davis puts it:

> modulating the activity of the brain during training or during sport will lead to benefits comparable to those of using drugs. The devices needed to generate these effects are already available, and are currently in use in laboratories or clinics to produce short- or long-term changes in performance. (Davis 2013, 649)

Another possible advantage to using brain stimulation is that the risks associated are relatively low as long as the technique is not abused. However, it is true that the information needed to set limits on brain stimulation or to know the long-term effects that it may have on athletes is in fact currently lacking.

The main brain stimulation techniques that are currently available are as follows:

(1) Transcranial magnetic stimulation (TMS) provokes the depolarization or hyperpolarization of neurons in the brain. TMS utilizes electromagnetic induction to induce weak electric currents in a rapidly changing magnetic field. Thus, certain activity is generated in specific or general parts of the brain, allowing brain functioning as well as the interconnections established within it to be studied. These effects take place in the stimulation phase for several dozen minutes and make the long-term reorganization of brain activity possible if stimulation is applied at regular intervals (Davis 2013, 649–650).

(2) Transcranial current brain stimulation (TCS) is a technique within which two variants can be distinguished; however, here we will focus on transcranial direct current brain stimulation (tDCS), which is characterized by a neural stimulation method that utilizes a constant low current, directly applied via small electrodes to the area of interest in the brain. The magnetic field's magnitude and polarity on the brain's

surface close to the electrodes determines its effect: the cells in the anode area increase in excitability through this process which involves modulating the resting cellular membrane potential. Initially developed as a therapy for patients with brain injuries, it has been shown to increase cognitive abilities in different ways depending on which part of the brain is stimulated (Kanai et al. 2008, 1839).

According to Davis, the most notable difference between the two techniques is that the former focuses on connected areas of the brain while the latter's effects are spread across the whole brain. Nevertheless, TCS offers the advantage of being more affordable and more portable. In fact, wireless TCS stimulators are already commercially available and there are websites that give instructions for home-made varieties of this device.

For this reason, Davis argues that brain stimulation will become the key technology in the future of sport and of sport medicine. There are two effects, according to this author, that neurodoping will have on sport performance. The first has to do with temporary performance. For a 20- to 60-min period following treatment, there was improved response time and time-to-fatigue as well as suppression of tremor. After a period of time, the effects declined but the usefulness this technique may have on athletic competition is irrefutable; take, for example, the usefulness of their effects on tests of speed or jumping.

The second use of this neuroenhancement is in the acquisition of skills:

> Skills learned in the context of anodal tDCS are acquired more rapidly, and reproduced more accurately, than those learned without. Sports performance at the highest levels require[s] good technique and good timing. These are skills learned during training, so enhancing the efficiency of learning during the training phase will be of greater benefit at competition time. I suggest that an athlete could use these techniques to make training more efficient and thereby gain an advantage. (Davis 2013, 652)

It is not difficult to predict, therefore, that neuroscience's application to training and enhancement in sport is only just beginning given that this powerful instrument is already in the hands of coaches and training staff. Both kinds of brain stimulation can have a great effect, for example, on archery by reducing tremor. They can also greatly impact tennis since the possibility of success is closely related to repeatedly winning the serve.[29] These are all trainable skills that can also be improved using these techniques. In any case, due to the great diversity of athletic competition, the acceptance and usefulness of these brain stimulation techniques may also be quite diverse; thus, any decision that is taken with regard to them ought to take the specific type of sport into account.

Nonetheless, transcranial stimulators are not exempt from problems or drawbacks. In the following, some of the most evident ones will be examined.

6. Some criticisms of cognitive enhancement and theories of sport

The first critical observation that must be made is that the effectiveness of these techniques has not been tested under real competitive sport conditions or for specific actions. In addition, as Davis points out, the tests have not been carried out on athletes but rather on normal people, that is, non-athletes. As a result, Davis is sceptical about applying them to elite sports.

A second objection would be that with neuroenhancement the athlete would not need to strive to make an effort or a sacrifice to obtain results. The acquisition of physical power, or other relevant skills for the sport would make the physical sacrifice of the sportsman irrelevant to obtaining the sport victory. But the impact of neurodoping in sport practice will likely not affect equality to such an extent. Neurodoping, at least in its current state, does not offer such miraculous effects that the athlete will obtain stratospheric results. The athlete still needs to train and make sacrifices to ensure top performance. In the end, neurodoping simply offers a small difference in the results; if an athlete were to rely on the miraculous effects of a pill or an electromagnetic session and stop training, it is highly unlikely that he would be in the elite of his field.

Another type of objection stems from dangers of potential generalized use, as well uncontrolled use by fans who are also athletes. The effects of doping in the gyms and athletic facilities of sport aficionados are well known, making it highly likely that the use of cognitive enhancers and cranial stimulators may also become widespread. Clearly this could bring about pressing public health concerns resulting from irreversible harm to the brain that could be caused by improper use.

A more general objection refers to cognitive enhancement. Certainly, any attempt to intervene in the brain can raise hotly debated issues such as authenticity and practical relevance when it comes time to judge the actions of an individual. There are others, such as McNamee and Loland, who believe that the imposition of paternalistic measures is in fact justified in that they keep sport from becoming the field for experiments in which athletes are used as human guinea pigs.

The first objection is that a possible side effect of this kind of enhancements is the creation of inauthentic personalities. The second objection argues that enhancers would change our mind in such a way that it will be difficult for us to attribute moral accountability. There are two possible answers to both criticisms. Firstly, it is unclear why the enhanced self is evaluated as inauthentic, especially if the enhancement is ongoing rather than momentary. The sport person who has inferior memory skills could negate these traits as the inauthentic by-product of a biological weakness. The authentic person is the one who fights against their imposed weak nature.[30]

Regarding the second objection, Kahane (2011) offers an interesting defence of enhancers as authentically coherent insofar as one tries to conform to one's own desires, preferences or values.[31] For these points of view, these techniques represent no more than advances over traditionally used techniques in education.

Such discussion warns of the possible consequences of enhancement for sporting purposes. Under the current anti-doping policy, athletes have the capacity to make informed choices about the use of enhancers in terms of their short- and long-term impacts: the possible enhancement (or not) of sport performance relative to a potentially reduced (or higher) level of welfare in future life.

Finally, what remains to be made known is the position taken by the World Anti-Doping Agency. Whether these substances should be included on the agency's famous list of prohibited substances is not a simple issue. Certainly, as we have just seen, the effects of these techniques on athlete performance are as of yet unknown; thus it would be risky to venture an opinion on that matter. But given their characteristics, these stimulation techniques still may receive identical treatment to hyperbaric or cryogenic chambers, whose effects are similar to those included on the WADA list although these treatments are not included in it. In any case, transcranial techniques would avoid the objections to technological unfair play given that their affordability would increase athletes' equal access opportunities.

It can be clearly seen that all of these matters are intertwined. It is also obvious that whatever decision is finally adopted by WADA regarding the use of these techniques must be based on some criteria that lend robustness to the argument. That is, the criteria should be morally based. Perhaps for that reason, it is advisable to finalize this article by turning to what Loland deems acceptable uses of technology, unacceptable uses of technology and uses that are of value to technology in sport.[32]

From our point of view, cognitive enhancement techniques open up an interesting field in terms of sport performance, an arena which demands close examination. In the first place, the lack of conclusive evidence regarding their efficacy under actual competitive conditions leads us to recommend not taking hasty decisions regarding their use. That is to say, they may or may not end up being of value to sport, but in order to know this, it is necessary to explore the possibilities.

Do not misunderstand: we are not defending a genuinely instrumentalist vision of sport.[33] Our position is somewhere between what Loland calls 'narrow and wide ideal-typical theories of sport'.[34] It is our opinion that sport should be open to the advances of scientific and technological knowledge. Among other things, this is because with them new frontiers to sport performance are opened up for exploration. And the Olympic motto, *citius, altius, fortius*, that is, progression in performance, seems to be an intrinsic part of modern sport. That does not mean that performance should be the true normative standard of sport. In reality, reliable evaluation of athletic performance can only be aptly understood when other intrinsic values of sport are also taken into account. In addition, it is not necessary to launch into a prolonged discussion regarding what these values are and what interpretations should be made based on them. Be what they may, they are not totally separate from sport performance, as claimed by the wide theory of sport that Loland defends and we partially endorse. Rather the opposite is true. Sport performance is a defining characteristic in highly competitive sport. It has been this way since the beginning and continues to be so in current times. Whatever the appropriate uses may be for technology, in order to attain them, it goes without saying that there is ample room for disagreement in this area. Whatever the disagreement, however broad it may be, to be adequately understood, it must be founded on some common ground, which in this case is none other than the link between the ends and the means. That is to say, the compatibility thesis.

7. Conclusions

As was seen at the beginning of this article, elite athletes are characterized by their pushing of the limits of physical and psychological performance. In fact, the exquisite combination of both is what is said by some authors to characterize the athletic identity or sport personality. But it is equally true that competitive sport has always made use of technology to constantly improve performance and in this way make good on the Olympic motto.

All of this leads to a recurring debate regarding the legitimacy of using enhancement techniques and substances in sport. In all likelihood, this debate has gone hand in hand with the evolution of sport itself. After all, the difference between the usage of natural substances with enhancement effects in the past and the current techniques of present times lies in the how and not the what. Consequently, it is unsurprising to catch glimpses of a near future in which there is increasing interest in biotechnological enhancements since these enhancements are likely to offer more potent and safer possibilities.

In current times, available techniques and substances not only allow improved physical abilities. The scientific evidence regarding the connection between the brain (cognitive and emotional factors) and the rest of the human body on sport performance requires that attention be paid to the enhancement of psychological abilities, and with them athletic performance, which is part of the scientific research agenda.

The second part of the article focused on the importance that cognitive abilities have on sport performance and in what way the extraordinary advances in knowledge about the brains can affect not only better comprehension but also better manipulation of the athlete's brain in order to improve their athletic performance. For this reason, the growing importance of neuroscience and its techniques as applied to the world of sport can be easily understood.

To get a general overview of these neuroscience techniques, we have analyzed the impact of two cognitive enhancers (modafinil and methylphenidate) and of transcranial stimulators on sport performance. As we have seen, transcranial stimulators allow neurons to be excited, directly influencing sport performance and thus improving time to fatigue or the ability of athletes to acquire new skills.

Next, we turned to the objections to using these sport performance enhancing techniques and their eventual inclusion on WADA's prohibited substance list. Furthermore, we echoed the potential danger that the widespread usage of these stimulators would have on athletes, given the fact that they are economically accessible. All of this does nothing but highlight the fact that this is an interesting and necessary debate that must be confronted by sport theory.

To conclude this article, we have highlighted some of the normative problems spawned by the use of these techniques in sport. We pointed out that given the lack of conclusive evidence for their efficacy in actual competitive situations, it would not be prudent to adopt definitive decisions with regard to them. We also maintained that, as occurs with any other brain intervention, they must be governed by meticulous respect for the principle of precaution in order to maintain the absence of undesirable harm to the primary organ of the human being. Lastly, we attempted to balance the arguments in favour of the advantages provided by easy access with those that warn against the possible risks to public health generated by inappropriate usage of the same.

In any case, all of these objections can explain and justify the adoption of restrictive measures in sport, but we must not forget that all of them refer to the technique, and as such, this objection will be surpassed by advances in the area. What will not vary is the normative question regarding compatibility with values of sport. Our position in this regard is moderately positive, or if you will, close to what Loland calls the 'narrow theory'. That is, we consider the compatibility thesis to be more hopeful because it allows high levels of performance to be attained without betraying the internal values of sport.

All said, given that this is a recurring debate in sport theory which has enormous practical relevance for athletes, it would be fitting to pay attention to scientific evidence and not hastily or contradictorily conclude, as the WADA list seems to do, that it is justified to ban some of these techniques while permitting others, such as hyperbaric chambers or cryogenics, which have identical effects.

Notes

1. McNamee (2008, 37).
2. Druzheyskaya et al. (2008).
3. http://whqlibdoc.who.int/publications/2010/9789241599979_eng.pdf.

4. Druzheyskaya et al. (2008).
5. Luthar et al. (2000, 435).
6. Morgan et al. (2013, 552).
7. Carrio (2015) Clearly, it is difficult to understand that individuals whose physical and psychological performance exceeds habitual norms of performance, but who are subject to accidents or illnesses just as any other person, may be extremely restricted in their ability to use of certain therapeutic substances. This restriction may virtually nullify the extraordinary characteristics of the individual's physiology.
8. Gaffney (2015).
9. Brewer et al. (1993)
10. Danish (1983)
11. 'Sport technology, then, are human-made means to reach human interests and goals in or related to sport' Loland (2009, 153).
12. Ibid.
13. Particularly what has been called 'technology administered by experts'; Loland (2009, 155)
14. Sceptics can always object that there is an ongoing debate in neuroscience regarding the mind-body duality. But it should be noted that the techniques of interest here, such as neuron stimulation, operate on the body part in this debate, i.e. the brain. As a result, the debate is becoming inconsequential in this regard.
15. Although we cannot detain ourselves on this point, it is worth differentiating between brain interventions that affect cognitive abilities from those that influence moods or emotional states.
16. Loland (2009, 153)
17. Ibid.
18. The case of the Australian skier, gold medallist in the winter Salt Lake Olympic Games, perfectly exemplifies the role of luck in competition. Certainly Bradbury qualified thanks to the disqualification of one of the favourites, Canadian Marc Gagnon. In the last turn of finals, the first four who all led Bradbury by more than 15 metres all fell, allowing Bradbury to be named the winner. It was the first Olympic metal won by a competitor from the southern hemisphere.
19. In some way, the distinction between different types of athletic competition is interesting because doping does not uniformly affect all sports. This is what S. Loland calls the 'thesis of vulnerability', which suggests that the essence of certain sports would be more greatly affected by doping than others. As we will see further on, something similar occurs with emotional doping.
20. Platonic and Cartesian dualism have been the dominant conceptions of the mind and body in sport, establishing the comparison between body and machine. Recently there has been noteworthy attention paid to the relation between both of these factors in sport performance. See Kretchmar (2013) and Illundáin (2013).
21. Some of these practices are imagery training and cognitive restructuring.

    Imagery training is the 'symbolic repetition of a physical activity in the absence of any large muscle movement'. The athlete methodically, consciously, and repeatedly imagines a sport action without actually physically executing it at the same time. The objectives of imagery training are: (a) facilitating motor skills to enhance practical execution; (b) controlling attention and concentration; and (c) accelerating recovery from injury.

    Cognitive restructuring is a 'set of techniques which intend to directly change the athletes' thoughts to better face the demands of competition'. The aim of this is to: (a) improve athletes' self-confidence; (b) strengthen motor activities; and (c) control attention and concentration. Vid. Sánchez and Lejeune (1999), 23.
22. To give an approximation of the term, neuroscience consists of the study of brain mechanisms which form the basis for an individual's essential cognitive functions: the ability to remember, argue, decide, etc. These functions can be observed through powerful instruments such as functional magnetic resolution imaging (fMRI), positron emission tomography (PET) and electroencephalography, which monitors the electrodynamic flow of neurons.
23. At the outset, it was used in medicine to treat attention deficit and hyperactivity disorder. Nevertheless, there is current widespread perception that it can have similar effects on people who do not suffer from hyperactivity or attention deficit. The consequence has been its spread

to university and secondary school students who have discovered that stimulants improve concentration. In this way they are being used not only for improvement in studies but also to improve exam performance. They have been shown to be used among university professors and researchers. Methylphenidate and modafinil are currently on the list of banned substances. Presidential resolution of 20 December 2013 Supreme Council for Sport, by which substances and methods are prohibited in sport. Dubljević and Ryan, 2015 'Cognitive enhancement with methylphenidate and modafinil: conceptual advances and societal implications', Neuroscience and Neuroeconomics: 4.
24. According to data published by NDCHealth, Adderall XR is on the list of the 200 most-sold medications in the United States. Specifically, it is number 69 and generated 730 million dollars in revenue for the last year. It is immediately followed by another stimulant Concerta (extended release methylphenidate) with sales of 270 million. Provigil (modafinil) generated 420 million in income, just below another drug that intuitively seems much more popular than the stimulants: fluoxetine (an antidepressant popularized in the 1990s under the trade name Prozac; available today generically) with sales of 450 million. Wikipedia: http://es.wikipedi-a.org/wiki/Estimulante. Last accessed 30/12/2013.
25. Reardon and Creado 2014, 98.
26. Peterson (2008).
27. According to the standard characterization, stimulants are drugs which increase levels of motor and cognitive activity, reinforce wakefulness, and the state of alertness and attention. The US Anti-Doping Agency defines a stimulant as 'An agent, especially a chemical agent such as caffeine, that temporarily arouses or accelerates physiological or organic activity'.
28. However, its inclusion on the list of prohibited substances is questionable to the extent that its enhancement effect is within 'normal' (not transhuman) limits and the risks it presents to health are not significant. It should be then asked whether its inclusion on the list of banned substances is justified.
29. Adjusting how the ball is struck and addressing it to a specific area of the tennis court.
30. This argument is a personal version of Foddy argument against the critics to mood enhancement based on inauthenticity:
    It seems highly plausible that an athlete would repudiate his tremor, or his nausea, or his perspiration in this manner. To tell him that his tremulous, sweaty, and nauseated self is his true self seems no more reasonable than telling dieters that it would be more authentic for them to remain overweight.
31. 'If authenticity involves being true to oneself, or to one's values, then there is a sense in which … when one uses mood enhancers, one is at most conforming to one's values.' (Kahane, 170).
32. Loland (2009, 156).
33. Russell (2005).
34. According to Loland there are three Ideal-typical theories of sport that express alternative normative interpretations of sport, namely, the 'relativist', the 'narrow' and the 'wide' theories. The narrow one embraces some technological optimism. Even if it could be, as Loland says, 'politically incorrect', the alternative understanding of the values of sport that it supports is 'clear and consistent' as Loland (2009, 156–157) himself recognizes. On the other hand, the wide theory is not blind to developments and innovations in sports. Even if it is critical to performance-enhancing expert- ad-ministrated technology, 'there are varying opinions among wide theorists about the justification of harmless variants such hypoxic tents' (Loland 2009, 158).

Acknowledgement

We thank the anonymous reviewers from Sport, Ethics and Philosophy for their careful reading of the article that help us to improve it.

Disclosure statement

No potential conflict of interest was reported by the authors.

ORCID

Alberto Carrio Sampedro http://orcid.org/0000-0002-8482-0190

References

BREWER, B.W., J.I. VAN RAALTE, and D.E. LINDER. 1993. Athletic identity: Hercules' muscles or Achilles heel? *International Journal of Sport Psychology* 24: 237–254.

CARRIO, A. 2015. When better is worse. On the therapy/enhancement distinction in sports. *Sport, Ethics and Philosophy* 9 (4): 413–426.

DANISH, STEVEN J. 1983. Musings about personal competence: The contributions of sport, health, and fitness. *American Journal of Community Psychology* 11 (3): 221–240.

DAVIS, N.J. 2013. Neurodoping: Brain stimulation as a performance-enhancing measure. *Sports Medicine* 43 (8): 649–653. doi:10.1007/s40279-013-0027-z.

DRUZHEVSKAYA, A.M., A. AHMETOV, and V.A. ROGOZKIN. 2008. "Association of the *ACTN3* R577X polymorphism with power athlete status", in Russians. *European Journal of Applied Physiology* 103: 631–634.

DUBLJEVIĆ, V. and C.J. RYAN. 2015. Cognitive enhancement with methylphenidate and modafinil: conceptual advances and societal implications. *Neuroscience and Neuroeconomics* 4: 25–33.

ERONIA, O. 2012. Doping mentale and concetto di salute: a possibile regolamentazione leg-islative. *Archivio penale* 3: 1–23.

FODDY, B. 2011. Enhancing skill. In *Enhancing human capacities*, edited by R. Ter Muelen, J. Savulescu, and G. Kahane. Oxford: Blackwell: 313–325.

GAFFNEY, P. 2015. The nature and meaning of teamwork. *Journal of the Philosophy of Sport* 42 (1): 1–22.

GOODALL, S., G. HOWATSON, L. ROMER, and E. ROSS. 2012. Transcranial magnetic stimulation in sport science: A commen-tary. *European Journal of Sport Science* 14 (Suppl 1): S332–S340.

HOBERMAN, J. 1992. *Mortal engines. The science of performance and the dehumanization of sport*. New York, NY: The Free Press.

ILLUNDÁIN, J. 2013. Moving wisdom. Explaining cognition through movement, Fair Play. *Revista de Filosofía, Ética y Derecho del Deporte* 1.

KAHANE, G. 2011. Reasson to Feel, Reasons to Take Pills. In *Enhancing human capacities*, edited by R. Ter Muelen, J. Savulescu, and G. Kahane. Oxford and Malden, MA: Blackwell: 166–178

KANAI, R., L. CHAIEB, A. ANTAL, V. WALSH, and W. PAULUS. 2008. Frequency-dependent electrical stimulation of the visual cortex. *Current Biology* 18 (23): 1839–1843.

KRETCHMAR, S. 2013. Mind-body holism, Paradigm and Education. *Fair Play. Revista de Filosofía, Ética y Derecho del Deporte* 8 (1): 28–43.

LEUNES, A. 2011. *Sport psychology*. London: Icon Books.

LUTHAR, S.S., D. CICCHETTI, and B. BECKER. 2000. The construct of resilience: A critical evaluation and guidelines for future work. *Child Development* 71 (3): 543–562.

LOLAND, S. 2009. The ethics of performance-enhancing technology in sport. *Journal of the Philosophy of Sport* 36 (2): 152–161.

MCNAMEE, M. 2008. *Sports, virtues and vices. Morality plays*. London: Rout-ledge.

MERKEL, R., G. BOER, J. FEGERT, T. GALERT, D. HARTMANN, B. NUTTIN, and S. ROSAHL. 2007. *Intervening in the brain. Changing psyche and society*. Berlin: Springer.

MORGAN, PAUL B.C., DAVID FLETCHER, and MUSTAFA SARKAR. 2013. Defining and Characterizing Team Resilience in Elite Sport. *Psychology of Sport and Exercise* 14 (4): 549–559. doi:10.1016/j.psychsport.2013.01.004.

PETERSON, D. 2008. Performance Enhancement Common in Sports. *Live Science*. Available at http://www.livescience.com/5230-performance-enhancement-common-sports.html (accessed 17 October 2016).

REARDON, C.L. and S. CREADO. 2014. Drug abuse in athletes. *Substance Abuse and Rehabilitation* 5: 95–105.

RUSSELL, J.S. 2005. The value of dangerous sport. *Journal of the Philosophy of Sport* 32 (1): 1–19.
SÁNCHEZ, X. and M. LEJEUNE. 1999. Práctica Mental y deporte. ¿Qué sabemos después de casi un siglo de investigación?. *Revista de psicología del deporte* 8 (1): 21–37.
SANDBERG, A. 2011. Cognition enhancement: Upgrading the brain. In *Enhancing human capacities*, edited by J. Savulescu, R. ter Meulen, and G. Kahane. Malden, MA; Wiley-Blackwell: 71–91.
TAMORRI, S. 2004. *Neurociencias y deporte. Psicología deportiva y procesos mentales del atleta*. Barcelona: Ed. Paidotribo.

ⓐ OPEN ACCESS

Skiing and its Discontents: Assessing the *Turist* Experience from a Psychoanalytical, a Neuroscientific and a Sport Philosophical Perspective

Hub Zwart ⓘ

ABSTRACT
This article addresses the question whether skiing as a nature sport enables practitioners to develop a rapport with nature, or rather estranges and insulates them from their mountainous ambiance. To address this question, I analyse a recent skiing movie (*Turist*, 2014) from a psychoanalytical perspective (skiing as a quest for self-knowledge and as therapy) and from a neuro-scientific perspective (ski resorts as laboratory settings for testing physical and psychic responses to a variety of cues). I conclude that Jean-Paul Sartre's classical but egocentric account of his skiing experiences disavows the technicity involved in contemporary skiing as a sportive practice for the affluent masses, which actually represents an urbanisation of the sublime, symptomatic for the current era (the anthropocene).

Introduction: Skiing and Philosophy

Skiing as a winter sport unleashes massive annual migrations of individuals, families and peer groups into mountain areas such as the Alps, notwithstanding the obvious drawbacks of this practice, both collectively (the disruptive environmental impact of ski resorts on valuable ecosystems) and individually (expenses involved in skiing, bodily injuries, etc.). Ski resorts endanger alpine ecosystems through continuous construction, enlargement and management of pistes with the help of machine-grading, grooming vehicles and artificial snow (Wipf et al. 2005). They are often established close to nature, in sensitive ecologic zones (Chappelet 2008), so that areas that were formerly deemed inaccessible are now increasingly exploited by the 'skiing industry' (Chivers 1994, 11), resulting in 'urbanisation' of Alpine areas (12).

Let me confess outright that I am a stranger, an outsider to skiing and that I have never participated in this pursuit. I prefer walking and hiking through the mountains, in a Rousseau-like fashion, avoiding ski resorts because from my perspective, whereas walking allows you to develop a rapport with nature, ski resorts rather seem to represent estrangement from the Alpine ambiance, by reducing the Alps (as an archetypically sublime landscape) to mere scenery. Skiing for me exemplifies forgetfulness of nature, to use the Heideggerian term,

---

This is an Open Access article distributed under the terms of the Creative Commons Attribution-NonCommercial-NoDerivatives License (http://creativecommons.org/licenses/by-nc-nd/4.0/), which permits non-commercial re-use, distribution, and reproduction in any medium, provided the original work is properly cited, and is not altered, transformed, or built upon in any way.

rather than interaction with or openness to nature. A ski resort is definitely a clearing, but *not* in the Heideggerian sense of the term. Besides the fact that ski resorts entail a devastation of the surrounding landscapes, they seem symptomatic of a commodification of nature, a demise of the sublime: of nature as φύσις: that which emerges, comes forward on its own accord, that which has its own inherent principles of movement and change, that which is simply there without our doing (Aristotle 1980).

At the same time, such outsider assessments should raise suspicion. As I never stood on skies myself, how can I *know* what skiing really means, or what kind of nature experiences (perhaps unknown to me) are opened-up by it? Moreover, my verdict may stand on the shoulders of a long tradition of philosophical suspicion and resistance against massive sportive entertainment. For although Plato's academy was situated in a sports park (offering gymnastics of the intellect in addition to athletic pursuits such as wrestling, discus throwing and running), authors like Seneca and Cicero already criticised the depraved morality of the public games (*circenses*) organised in ancient Rome (Sloterdijk 2001, 309), moved by sound philosophical reasoning no doubt (given the outrageous 'bestiality' of the human and animal suffering involved), but in the case of skiing I would like to probe and articulate my discontent with more precision. How to achieve this without becoming actually involved in this questionable practice myself?

It cannot be denied, moreover, that even in the case of skiing, there is a *sed contra est*, in the form of prominent philosophers who stand out as advocates of this practice. Perhaps the most famous philosophical ski adept was Martin Heidegger. Besides the fact that several pictures of Heidegger on skies can be found on the Internet, it is well known that one of his key concepts, namely *Die Kehre*, Heidegger's famous 'turn', is actually a skiing term, borrowed from the skiing vocabulary, Heidegger's preferred sport (Janicaud 2015, 242). Apparently, he was quite proficient in it, skiing all the way from his famous hut in Todtnauberg to the lecturing hall at the University of Freiburg (Breivik 2011; Moran 2000). The meaning of skiing may have changed dramatically during recent decades, however, so that nowadays skiing not only exemplifies forgetfulness of nature but also the inauthenticity of contemporary consumerist existence, with tourists buying commodified experiences and quality time (Pine and Gilmore 1999) instead of really exposing themselves to authentic ecstatic peak experiences of self-forgetfulness (Martínková 2011; Vannatta 2008).

A very explicit philosophical advocate of skiing was Heidegger's contemporary Jean-Paul Sartre who, in *Being and Nothingness* (1943), devotes almost a whole chapter to this pursuit. In the winter of 1934–1935, while struggling with his novel *Nausea*, Simone de Beauvoir persuaded Sartre to go skiing in the Alps near Chamonix together (Martin 2010; Pitt 2013) and Sartre's phenomenological first-person analyses in *Being and Nothingness* are based on these personal experiences. Sartre discusses skiing in a chapter dealing with *doing*, *having* and *taking possession*, important ingredients of his 'existential psychoanalysis', a phrase which indicates that, whereas Freudian psychoanalysis emphasises the extent to which our self-image, our whole worldview even, relies on self-deceit, Sartre wants individuals to assume full responsibility for their existence.

Before turning to skiing, Sartre first discusses scientific research as a form of 'appropriation' of nature. Scientific research, according to Sartre, means that certain aspects of the world reveal themselves via *me*. Nature, Sartre argues, is Diana taking a bath, and the researcher is the hunter who is suddenly and accidentally confronted with her whiteness, her nudity (a dynamics which Sartre refers to as the Acteon complex, 639). The basic desire of science is to take possession of this smooth, white body, which remains a hazardous task, but the researcher somehow has to conquer the *noema* of his desire. Thus, the goal of scientific research is to penetrate and incorporate the object: that which initially appears to me as decidedly *not*-me.

Besides penetration, there is another form of appropriation, namely contemplation, which is more like caressing and glancing over a surface, and it is here that skiing comes into play (640). Skiing as a sport, Sartre argues, transforms a natural environment into a supporting, enabling element for human action (642), and this explicitly applies to the Alpine slopes used by skiers: the white immensities which, according to Sartre, initially represent pure exteriority, radical spatiality and monotonous indifference: the absolute nudity of substance, the pure apparition of not-me (642), which the skiing subject desires to appropriate and possess. There is something ungraspable about snow, however, for it lacks substance. As soon as we touch it, it disappears, liquefies and evaporates into nothingness. But skiing allows us to *own* this element. The meaning of skiing resides in the experience of speed and technical ability no doubt, but first and foremost the skier desires to possess these beds of snow. For Sartre, sliding is a form of appropriation, and precisely for that reason, skiing is *not* a superficial activity. Sliding is a very special mode of appropriation, moreover, enabling a strictly individual relationship with matter. The ideal form of sliding leaves no traces at all, although in reality it always does, so that even Alpine skiing is already less than perfect, because there always is a visible trace, which compromises skiers (hence the disappointment which seizes them when they look behind them and see the imprint which their skis have left on the snow). Caressing a 'female' body whose skin we leave intact: that, for Sartre, is the skier's ideal. Through the activity of sliding, the skier affirms his rights over the snow: this is *his* bed of snow. But appropriation requires a kind of test: it means overcoming a difficulty, a certain resistance, such as: climbing the mountain before descending it, because skiers first have to conquer the smoothness and whiteness before they can appropriate it during the descent. Thus, for Sartre, skiing is a form of appropriation which realises a rapport with Being.

Even if we ignore Sartre's stereotypical depiction of the gender dimension (male subjects conquering, appropriating and owning white, smooth female bodies), his philosophical advocacy of skiing becomes quite problematic in view of the current era of ecological disruption, notably his claim that this act of conquering appropriation may result in establishing a *rapport* with nature. Would this not rather require an attitude of non-appropriation, of letting nature (letting otherness) be? In view of the massive melting of Arctic ice and the shrinking of the Alpine glaciers, Sartre's depiction of the Alps seems remarkably outdated and (even in the Sartrean sense of the term) irresponsible: a symptom of environmental self-deceit. Overall, Sartre's *apologia* of skiing reinforces rather than refutes my intuitive assessment. On the positive side, there is (a) the experience of skiing as a form of exploration, an instantiation of our desire to know nature (representing a common ground between practising sports and conducting research); (b) the idea of skiing as a test and (c) skiing as an instantiation of the desire to build a rapport with nature via physical activities, notably the activity of sliding which, ideally, leaves no traces. On the negative side, however, there is first of all the desire to appropriate nature and to transform it into a lusory ambiance. But what renders Sartre's existential psychoanalysis of skiing even more inadequate and unsatisfactory (as a potential contribution to the contemporary philosophical debate on sports and the environment) is his depiction of skiing as a decidedly solitary and individual pursuit.

What is completely eclipsed in his egocentric portrayal is that skiing requires an infrastructure, a material base: a network of hotels, roads, ski lifts, snow cannons, snow-grooming vehicles and so on. The technicity of skiing, everything which has to be in place in order to transform Alpine nature into a supporting ambiance for the skiing industry, fades into the backdrop, and this makes Sartre's portrayal fairly 'superficial'. Doubtlessly, in the 1930s, skiing had not yet evolved into the massive tourist attraction which it has become today, so that in those days it could perhaps still be considered a 'nature sport' (whose practitioners interacted with natural features: Krein 2014, 2015), but insofar as skiing during the interbellum really was a solitary, low-tech activity, this reduces the relevance of Sartre's account for the current sport philosophical debate because what used to be marginal and rudimentary perhaps (the technical, infrastructural side of skiing) has now become decidedly visible and dominant. And my negative attitude towards skiing does not concern the physical activity *as such* of course, but rather everything that comes with it; and precisely this is overlooked in Sartre's self-centred account, which focuses on the experience of *sliding*.

Thus, Sartre's existential psychoanalysis of skiing results in an aporia, a deadlock. Somehow, I must find another way to put my intuitive verdict (my prejudice?) to the test. According to Breivik (2011, 319), we should only speak about sports on the basis of personal experience. Otherwise, we are unable to study aspects that only reveal themselves to those who have reached a sufficient level of mastery. But if *practising* skiing is out of the question (for environmental reasons), what could count as a convincing substitute? My strategy on such occasions is to revert to genres of the imagination (novels, movies, drama, art, etc.) that open up feasible, in-depth perspectives on practices in which I am not directly involved myself. Several authors have pointed out, moreover, that there is a dramatic side to sport, a resemblance between genres of the imagination and sport activities (Gebauer 1993; Kretchmar 2017; McNamee 2007). In both cases, there is a beginning, a confrontation, a potential rise of tension, a climax (or crisis) and a resolution (Gottschall 2012). Keenan (1973) described sport as a 'tragic form of art' in terms of plot, character development and dramatic structure. More specifically, where movies are concerned, Crosson (2013, 3) has suggested that it is no coincidence that modern sport and cinema emerged contemporaneously, towards the end of the nineteenth century. In 1896, when the first Olympic Games were held in Athens, the Lumière Brothers demonstrated their *cinématographe* in Paris (3). Indeed, cinematic documents, from *Olympia* (Leni Riefenstahl's classic documentary of the 1936 Berlin Olympics) up to *Million Dollar Baby* (2004), seem to exemplify this congruency, and in the Internet Movie Database (IMDb) something like 2000 sport movies are now listed.

In order to further explore my intuitions, I opted for a recent Swedish movie named *Turist*, directed by Ruben Östlund (2014) and translated as *Force Majeur*, which provides a cinematic window, allowing me to critically reflect on skiing as a sport. In this article, I will use the movie setting provided by *Turist* as a philosophical laboratory, where various aspects of skiing can be analysed and assessed. In *Turist*, a ski resort creates an artificial environment where individuals and families (as 'research subjects') are exposed to various challenges and circumstances, both mentally and physically, and the question is: How will they respond? This is precisely what we see happening in the movie. The protagonists are confronted with a completely unexpected and threatening (test-like) situation, to which they must react in a split of a second, resulting in self-knowledge. While analysing the movie from multiple perspectives (psychoanalysis, neuroscience and continental philosophy of sport), special attention will be given to the methodological aspects of the use of cinematic documents as 'empirical' input for normative deliberations.

I will start with a brief resume of the narrative. Subsequently, I will analyse the movie by mutually exposing a psychoanalytical with a neuroscientific perspective, a procedure referred to as triangulation. Finally, I will assess what we have learned about skiing from this case study, and how the dissection of movies may contribute to the philosophy of sport.

Summary

An apparently happy and affluent Swedish family (the couple Tomas and Ebba, together with their children Vera and Harry) arrives in an expensive ski resort for a five-day ski vacation in the French Alps. The landscape is ostensibly managed, which involves a considerable amount of technicity. Denotations occasionally resound, coming from snow cannons used for avalanche management, interrupting the (usually rather brief) conversations as well as the background music (Vivaldi). The silent white scenery of the gigantic Alps in the distance contrasts with the family's everyday routines (brushing teeth, visiting the toilet, waiting for the lift, checking tablets or iPhones, playing with a drone, eating, drinking, ordering drinks, arguing, sleeping). While having lunch on a terrace with some visitors, a detonation suddenly resounds and an avalanche comes down from above, but father Tomas reassures his children that everything is under control. But as the avalanche comes increasingly close, a general panic ensues. In a split of a second, Tomas grabs his gloves and iPhone and dashes off, leaving his wife and children behind. As the sudden fog gradually disappears, he returns to the terrace, now covered under a film of snowy powder, but it is clear that something has changed. By deserting them, he has shocked and disappointed not only his family members, but also himself. Initially he denies his cowardice, insisting that he did not run away from the table at all, a strategy which infuriates Ebba, who refuses to let it pass. His iPhone has recorded his embarrassing retreat, and when Ebba (during a conversation with Mats and Fanni, a befriended couple) confronts him with the evidence, his resistance collapses. It is now clear that the traumatic scene requires some serious working through on the part of the married couple. Somehow, Tomas's injured self-image must be repaired.

Exposition and Exposure

How to dissect this cinematic document as a sport philosophical *Fallgeschichte*? I will approach the movie from two perspectives, resulting in an exercise in philosophical triangulation (Zwart 2016). On the one hand, I will approach the movie from a psychoanalytical perspective, focusing on the experience of trauma (the sudden resurgence of something which seemed forgotten and disavowed) and the therapeutic sessions resulting from this. On the other hand, I will analyse the movie in terms of its experimental design: the movie as a lab-like personality test. In combination, both approaches provide input for a philosophical assessment.

To begin with, a number of dramatic stages can be distinguished in the movie, starting with the 'exposition' stage (Freytag 1863). Via the protagonists, we are introduced to the details of the ambiance: hotel rooms, restaurants, lifts, bedrooms and so on. Nature is visible as scenery, as the scenic backdrop, and the first shots seem to confirm that ski resorts basically entail an obliteration of the sublime. Tourists are not really *exposed* to or challenged by nature. They see the snow (the white and silent vastness) no doubt, while eating lunch or having a beer, but they do not really seem to *experience* the phenomenon. The rooms in which they sleep, the cabins that transport them safely to the top, the big machines that smoothen the slopes, the fashionable outfit they wear, the food they consume, everything seems perfectly designed to prevent them from any real confrontation with nature. Skiing apparently does not allow them to *experience* the environment, precisely because it is *not* at all a 'wild place', free from human influence (Dougherty 2007; Zimmermann & Saura 2017). The artificial scenery rather exemplifies estrangement from nature; they are barred from experiencing the Alpine 'real', to use the Lacanian term. This is the first moment in the dialectical unfolding of events: the control condition. While having their lunch on the terrace, the family members' attitude suggests complete indifference to the astonishing spatiality of the Alps.

But then, suddenly, there is a situation of high intensity, ambiguity and contradiction, a sudden exposure to the disregarded landscape and its looming threats (never completely domesticated): an 'intrusion of the real' (Zwart 2015), the sudden onset of the avalanche, which initially seems under control, but which gradually seems to get *out of* control, posing a direct threat to the nuclear family's survival.[1] In Lacanian terms, the real is that which offers resistance to complete 'assimilation' (Lacan 1964/1973, 65), revealing itself as a crevice, a rupture, as something totally unexpected or unacknowledged (Lacan 1960–1961/2001, 58). The real is a disruption which flouts our expectations, something which, from the point of view of everyday behaviour, seemed utterly 'impossible' (Lacan 1971–1972/2011, 141). As Heraclites phrased it many centuries ago: real nature is wont to hide herself and a sudden revelatory experience may prove quite disconcerting or even traumatic (Lacan 1962–1963/2004, 85 *ff.*). In other words, that which seemed contained suddenly comes to the fore, or *dys*-appears (Leder 1990; Zeiler 2010); although in this case what was absent is not the absent body but rather absent, disregarded nature.

Yet, at the same time this experience seems highly *unreal*. In accordance with the experimental set-up, this sudden, unexpected, unpredictable and life-threatening event is actually a phantom crisis, subjecting the protagonists to a kind of test: How will they respond? In *Turist*, skiing is not a *contest* (neither with others, nor with nature: Kretchmar 1975) but rather a *test*, and this especially applies to Tomas. The avalanche confronts him with a question: What kind of person (man, father, partner, etc.) are you? That the setting works like a test is confirmed by the multiple discussions that evolve during the remainder of the movie. Mats and Fannie for instance run into a discussion over the issue how Mats himself would have responded. But this question is also taken up by audiences: What would *you* do? How would *you* respond?

The Experimental Movie

Seeing movies as a test-like, experimental design builds on a concept developed by the French novelist Émile Zola (1840–1902), namely the concept of the 'experimental novel', which is here extrapolated to movies. After reading Claude Bernard's famous introduction to experimental medicine (Bernard 1865/1966), Zola (1880/1923) concluded that novels are basically laboratories and that novel-writing means conducting an experiment, by consciously and systematically exposing protagonists (with their personality traits, cultural backgrounds, professions, age, gender, etc.) to experimental stimuli and conditions in order to register and analyse their responses (Conti and Conti 2003; Zwart 2008, 2014). The novelist does not merely *describe* the world as it presents itself to us, Zola argues, but actively *intervenes*, confronting literary characters with a number of challenges introduced by the author.

Thus, novels (but this also applies to movies, I argue) may contribute to our body of knowledge. Indeed, according to Zola, the experimental method puts novel-writing on a scientific footing, so that novels may function as imaginative laboratories, where social phenomena may be systematically analysed. To achieve this, the experimental novel must convey the same sense of detachment and precision as scientific research reports.

A perfect exemplification of the experimental novel, in Zola's sense of the term, is Jack London's *The Call of the Wild* (1903/1981), which meticulously describes how a dog (named Buck), who grew up in sunny California, is kidnapped and shipped to Alaska to serve as a sledge dog during the Klondike Gold Rush, where he finds himself exposed to completely different conditions: the cold and frost of the Arctic North, the ghostly calm of its white and silent vastness, as London phrases it. He finds himself 'jerked from the heart of civilization and flung into the heart of things primordial' (30), resulting in a 'retrogression' (38), so that his 'civilized self' is gradually erased. Notably, the Californian ethic of love and fairness gives way to the laws of primordial life, the struggle for survival. His civilised canine morality disintegrates and he develops new behavioural repertoires, new habits of sleeping and eating, new senses even. Prolonged exposure to harsh circumstances entails a complete metamorphosis.

If we watch *Turist* from this perspective, the ski resort (an artificial location in an unfamiliar environment) becomes a laboratory setting. Initially (in contrast to *The Call of the Wild*), the sheltered, comfortable setting seems like an urban capsule transplanted into a mountainous region. There is something highly artificial about this insulated environment. Ski-lifts, terraces, hotel rooms, restaurants and everything in it seem like laboratory props, brought in by the research team to optimise the ambiance. The avalanche (produced on purpose with the help of snow cannons) becomes a 'stimulus', an experimental condition: a brief exposure to the *wild* whose *call* had been submersed. How will the protagonists (the research subjects) cope with this unexpected situation, how will they respond?

The interesting thing is: you cannot know this beforehand. The movie (the avalanche scene) confronts us with something we do not know about ourselves, unless we are willing to expose ourselves to such a situation, to such a test, as naïve subjects preferably, in the experimental sense of the term: subjects who do not *know* that they are actually being subjected to an experiment. The avalanche is a convincing scene, I would argue, because the situation really seems to work as a plausible personality test.

In fact, such test-like experiments are really being conducted in laboratories across the globe. And the outcome consistently is that if we are confronted with a sudden, seemingly life-threatening stimulus, three types of actions, three behavioural options, may ensue: fight, flight or freeze—a behavioural repertoire we actually share with other mammals. Which of the options will prove to be the most optimal depends on the situation. In the case of a gazelle who suddenly finds itself face-to-face with a hungry lioness, freezing offers the best chance of survival. If such an animal manages to stand completely still, the predator might ignore it. But if you are a passenger on a cruise ship which suddenly capsizes, freezing is by far the worst response imaginable because besides significantly increasing your own risk of drowning, others ('fighters') may feel obliged to try and save you (thereby putting their own lives at risk, for instance by stumbling over other freezers they encounter in corridors, etc.). Flight seems a sensible response in such a situation, unless of course you are the Captain of the ship, of the Costa Concordia for instance, for in that case to desert your ship (while your passengers are still on board) will mean the end of your career and the beginning of troublesome legal procedures. But again, the quintessence is that *you do not know*, unless you have been tested. And it certainly makes sense, on some occasions, to put individuals to a test, to test their 'metal' as it were, before they decide to become a cruise boat Captain, for instance, or a stewardess or a police officer—or before selecting them as marriage partners.

Perhaps this is why some athletes become extremophiles, eager to expose themselves to extreme conditions: they want to put themselves to the test, they want to know something about themselves which otherwise would remain hidden. Dangerous sports put people from relatively safe societies into situations which highlight 'human finitude' as a fundamental component of authentic understanding (Martínková & Jim Parry 2016). Disruption of the everyday life world can provide a glimpse of authenticity (Breivik 2010, 2011). Sartre already underscored the continuity between scientific research (the quest for knowledge) and skiing, and this seems to be confirmed by *Turist*, although skiing seems directed at deepening self-knowledge rather than nature knowledge. Indeed, outdoor sports may paradoxically learn us more about ourselves than about the environment (Atherton 2007).

Coping with Anxiety: The Socio-Technological World as Behavioural Scaffold

As indicated, the fight–flight–freeze response is an important topic of research in the contemporary neurosciences, as exemplified for instance by the inaugural lecture of brain scientist/neuropsychologist Karin Roelofs entitled *Brain in flight: The anxious brain in action* (Roelofs 2012). In 2013, she and her group launched the *Neuro-defence* research programme, funded by the European Research Council and aimed at investigating freeze–fight–flight reactions in humans by focusing on the neural and endocrine mechanisms underlying these defensive responses (a line of research initiated by Walter Cannon in 1920). In her inaugural lecture, Roelofs refers for instance to the disaster which befell the ferry Estonia during a storm in 1994, an event which caused the death of 852 people, while only 137 passengers and crew members survived. By analysing witness reports, Leach (2004) had already discovered that while some victims (15%) acted in an efficient and goal-directed manner, the vast majority (70%) panicked, in other words: they reacted, but not in a very efficient and sensible way. The final 15% (most of whom drowned) reacted by freezing in the face of danger. They seemed paralysed by the threat, apparently in shock, even when a more active response could have saved them. Such reports raise the question: How would *I* respond. To find the answer, Roelofs designed a laboratory experiment, involving a 'stabilometric force platform' (Roelofs et al. 2010; Roelofs 2017) which exposes subjects to a threatening, a neutral or an affiliative cue, allowing them to test their metal, so that, when things get really rough, their behavioural impulses will not come to them as an (embarrassing?) surprise.

My argument would be that in *Turist*, the terrace facing the Alps (neutral or even affiliative at first, but threatening all of a sudden) functions as an experimental 'platform', comparable to the stabilometric force platform which Roelofs and her colleagues use to measure spontaneous bodily responses—albeit in an outdoor, less controlled ambiance in the *Turist* version. The ski resort seems a highly artificial environment, as most of the human–nature interaction is fairly superficial and the protagonists seem isolated from their world rather than interwoven and enmeshed in it. In terms of poise, they seem focused on the horizontal dimension (interpersonal relationships, equipment, meals, etc.) rather than on the vertical dimension: nature and its marvellous summits and distant peaks (Breivik 2011; Todes 2001).

They do not really seem to experience Alpine grandness and nature does not genuinely seem to *speak* to them, which makes the terrace a perfect site for a laboratory. The phantom crisis forces the *parasympathetic* Autonomic Nervous System (ANS)—i.e. that part of the brain which enables a 'rest and digest' physiology—to switch to a fight, flight or freeze response, unleashed by the *sympathetic* ANS, while switching off the parasympathetic 'rest and digest' parts of the system. In Tomas's case, this results in an embarrassing flight response. There is also a link with masculinity, moreover, because someone's tendency to act is a function of testosterone levels, so that administering testosterone may modify the type of response (Roelofs et al. 2010).

In recent decades, neuroscience further elucidated the cellular and molecular mechanisms underlying anxiety, arousal and the fight–flight–freeze response. Besides adrenaline, noradrenaline and steroid cortisol, the protein stathmin, encoded by the stathmin gene (STMN1), plays a crucial role. Stathmin knockout mice are significantly less sensitive to fear (both innate and learned) and show less anxiety when exposed to new and potentially dangerous situations (Brocke et al. 2010; Martel et al. 2008). Top athletes from the mythic past like Siegfried (who knew no fear) may now be seen as cases of stathmin deficiency, causing the Siegfried complex as it were, as the STMN1 genotype has functional relevance for the acquisition and expression of basic fear and anxiety responses also in humans (Brocke et al. 2010). Brain research has entered sport science as well, as exemplified by a recent book entitled *The Athletic Brain: How Neuroscience is Revolutionising Sport and Can Help You Perform Better* by Amit Katwala (2016), focusing on the plasticity of the brain, however: on the brain's ability to learn and on the ways in which sport changes your brain. The athletic brain is a trained brain as sporting skills become automated, so that athletes who are 'in the zone' experience a state of flow (optimal relaxed arousal).

Genealogically speaking, this type of research descends from the work of Ivan Pavlov (a key source of inspiration for Walter Cannon as well, Smith 2008) and his famous distinction between conditioned and unconditioned reflexes. Whereas unconditioned fear may be seen as an evolutionary relic, the hard-wired fright response can be subdued by learning: by establishing conditioned responses, for instance by habituating skiers to experiences such as speedily sliding down a slope or facing snowy white expanses. Conditioning may be seen as a form of compensation, psychoanalytically speaking. The primordial experience of anxiety (invoked by mountainous landscapes) is mitigated.

But neuroscientific research may easily slide into neuro-centrism, i.e. the idea that we, instead of being mundane and worldly beings (embedded in socio-cultural environments), basically *are* our brain. The avalanche experience in *Turist* is an antidote against this tendency. In normal, everyday situations, Tomas has learned to compensate and conceal his susceptibility to the flight response. But the avalanche reveals the extent to which his casualness actually depends on the socio-technical ambiance. When the reliable world suddenly and momentarily collapses, the conditioned responses become disrupted and the primordial response takes over. The avalanche is reminiscent, I would argue, of the famous disruptive event which struck Pavlov's laboratory on 22 September 1924 when, during a flooding of the river Neva, his conditioned dogs were forced to swim to the top of their cells to avoid drowning, until one of the laboratory assistants finally came to their rescue. As a consequence, in a significant number of dogs, the carefully established conditioned reflexes were completely deranged (Pavlov 1951). Their normal environment had worked as a scaffold supporting their learned behaviour, but now they were thrown back on their resurging, primordial, unconditional reflexes. Laboratories are artificial worlds designed to keep the chaotic complexities of the outside world at bay, and the flooding of the River Neva was an intrusion of the Real.

Therapy, Denouement and Working Through

The remainder of the movie is basically a painful dissection of Tomas's personality, revolving around his fatal mistake, his ἁμαρτία in the Aristotelian sense of the term (Aristotle 1982). It was a mistake, a *Fehlleistung* no doubt, but as Freud argued (1917/1940) it is via our mistakes (rather than via our athletic achievements, e.g. Tomas' technical ability to slide smoothly across the slopes) that we really get to know ourselves.

My thesis is that Tomas and Ebba, unconsciously perhaps, asked themselves the very question that is triggered by the movie: How will I respond? Tourism is a present-day version of a pilgrimage to the Temple of Apollo at Delphi in ancient Greece, undertaken with the objective of *knowing yourself*, as indicated by the famous adage written on the sanctuary's façade (γνῶθι σεαυτόν), enabling a 'turn' on the personal level, a μετάνοια (Welters 2016). Skiing in the Alps provides a lusory alternative to going on a pilgrimage or paying a visit to the neuro-defence laboratory, but on all three occasions, the motivating question is basically the same. The lunch on the terrace constitutes a quasi-experimental situation, exposing the family members to a threatening cue, with no real physical danger involved: a phantom crisis. They are subjected (semi-voluntarily) to a personality test. Question: Can your (can his?) instincts be trusted?

Tomas unfortunately fails the test. His response seems to give something away about himself which he apparently did not know himself. His own life, his iPhone even, has more value to him than the family he is supposed to protect. Psychoanalytically speaking, his role as father figure is based on self-deceit and his *Fehlleistung* betrays him, resulting in an embarrassing, yet ambiguous situation. For yes, he failed, in a normative sense of the term, but at the same time he is not responsible for his action because, from a neuroscientific perspective, it was his brain system which unleashed the unconditional response. His body reacted like a cybernetic machine. Remember that it all happened in a flash of a second. Tomas *himself* is a witness of his own behaviour. He reacts to his reaction in retrospect, but neuroscientifically speaking he was unable to prevent it. It was his amygdala, so to speak, sending brainwaves to brainstem and striatum, several milliseconds before his frontal neocortex realised what he was doing. Faced with an unexpected threat, he proved unable to inhibit his 'automatic action tendencies' so as to 'optimize adequate response capacity' (Roelofs 2017), but this is because, neuroscientifically speaking, emotional brain activity ('hot cognition') is able to process information in no time (within 20 ms or so), therefore responding much faster than rational brain systems ('cold cognition').

His initial response to Ebba's comments and questions is denial. Tomas self-edits his memories and flatly refutes the accusation, which infuriates Ebba even more. But post-modern existence creates panoptic situations because we are *always being seen*. Our behaviour is continuously recorded, as also happens under laboratory conditions (all the world's a laboratory nowadays). We are permanently under the gaze of the other, via self-tracking devices such as Tomas's iPhone, switched on to videotape the phantom avalanche. The very item he valued so much in the turmoil gives him away.

When Ebba presses Tomas to explain himself in front of the befriended couple, Mats supports his friend via 'rationalisation'. Maybe it had been a Darwinian response: get away immediately, so that you will be able dig out and safe your family members later. This interpretation seems implausible and does not really help. Working through and acting out seem the only options. And indeed, several therapeutic sessions emerge during the remainder of the movie (the avalanche already occurs on day two of their visit). At a certain point, Ebba confronts Tomas in the corridor. He cries and confesses how he hates himself, and she drags him back into their suite where he collapses, so that his children come to hug their sobbing father, urging Ebba to join the therapeutic hug.

Friend Mats tries another option by taking him off piste to a spot where they really seem to experience the immense snowy surroundings for the first time. Here, Mats advises Tomas to scream. Apparently, he had been in scream therapy himself and it had worked. And indeed, Tomas's scream echoes across the Alps. It is as if a rapport, a dialogue with nature is now established after all. Back in the resort, he is overrun by a large group of young naked males who seem to enjoy a collective, intimidating scream ritual.

But these sessions cannot eradicate the fact that all the family members (Ebba, their children, but also Tomas himself) now look at him with different eyes. Tomas's response to the threat entails a narcissistic offence. His self-image is damaged (a psychic variant of the physical damage which skiing so often has in stall for tourists). The disruptive experience also involves a gender dimension. Whereas Ebba 'instinctively' tries to save her children, Tomas's masculinity fails him and the collapse of his male identity is presented as an emasculated experience, psychoanalytically speaking.

Most of the working though entails sessions of the talking cure type, reminiscent of another Swedish movie: Ingmar Bergman's *Scenes from a Marriage* (1973). In the Nordic tradition of Ibsen, what has been experienced must be talked through, and other characters are dragged into the verbalisation of their marriage crisis. The objective is reparation and individuation. The traumatic disclosure must be embedded in Tomas's self-narrative again.

Philosophical Reflections: Triangulation

What does this cinematic case study contribute to the philosophy of skiing? From an existentialist perspective, *Turist* reveals how ski resorts as infrastructures transform a natural environment into an ambiance for lusory activity. What seems absent in Sartre's portrayal is quite ostensibly brought to the fore: Alpine nature framed as a *Gestell*, to use the Heideggerian term, so that the threatening naturalness and otherness of nature seems significantly subdued. The protagonists become trapped in an inauthentic situation, into which they have 'fallen' in the Heideggerian sense. Quite unexpectedly, due to the approach of the avalanche, which seems to get out of control, this protective technological screen is temporarily lifted or suspended, so that the family members suddenly find themselves face-to-face with the threatening real. This experience reveals that, contrary to Sartre's claim, the sense of mastery entailed in the experience of sliding over a slope is illusory. For all the technicity involved, nature is still *there*, and the skiers never really conquer it. In Sartre's own terms, his account represents an egocentric instance of self-deceit, resulting from a failure to acknowledge the level of technicity (environmental disruption) that is required, especially today, in the era of mass tourism, to facilitate the sliding experience described by Sartre: the experience of skiers that they allegedly *master* and *own* the Alpine slopes simply by sliding over them. In *Turist*, shots of people sliding downwards are actually quite rare, as most of the movie is devoted to a plethora of other activities entailed in visiting ski resorts. But the confrontation with the avalanche disrupts this self-deceit. It reveals the vulnerability of human beings (now momentarily thrown back onto their primal impulses, their body wisdom), but also the vulnerability of nature because there is something drastically artificial about these carefully managed avalanches, including the one that *seemingly* gets out of control. Even the phantom crisis ends as a melting film of harmless, snowy powder: something bereft of substance.

As a cinematic *Fallgeschichte*, we analysed *Turist* from multiple perspectives: philosophical 'triangulation' (Zwart 2016). The movie was approached from a Sartrean (existential psychoanalytical), a Lacanian (psychoanalytical) and a neuroscientific perspective (building on the concept of the experimental movie). Although *Turist* is not really an experiment (the terrace is not as abundantly equipped with technical contrivances as the Radboud stabilometric force platform for instance), its narrative structure nonetheless reflects an experimental design. This combined reading of the cinematic narrative (in terms of a psychoanalytic assessment and a neuroscientific test) allows us to discern the 'morale' of the story, its relevance for contemporary debates in sport philosophy.

The most important sport philosophical issue for me is the question whether skiing enables or prevents practitioners from establishing a genuine rapport with nature (the natural environment, the Alpine landscape). Will nature sports like skiing contribute to ecological disruption, or rather to an ecocentric turn, so that skiing may become nature-friendly and eco-compatible again? My (provisional) conclusion (based on a single cinematic document, $N = 1$) is that *Turist* emphasises what is obscured in Sartre's analysis, namely that the technicity of skiing (as a collective practice) reduces nature to scenery, a mere condition for lusory human activity (in accordance with Heidegger's concept of the appropriation, exploitation and framing of nature as a standing reserve, a *Gestell*). Sartre's existential psychoanalytical interpretation underscores that skiing is basically an egocentric practice which, rather than establishing a genuine rapport with nature, produces an illusory experience of mastery over nature (an illusion which is effectively demolished by the avalanche experience). In other words, the experience of conquering nature (by sliding gently across snowy slopes) is based on self-deceit. Alpine nature is conquered, *not* by the sliding individual, but by technological infrastructures: the framing of Alpine nature as an ambiance for human lusory activity.

But Alpine nature is never *completely* owned or conquered, and the mock avalanche is, philosophically speaking, a therapeutic event which allows us to rectify this deceptive view (framing ourselves as omnipotent beings while we actually remain fundamentally vulnerable and dependent on technology). Tomas's response (initially disavowing the threat, claiming that everything is under control, but subsequently opting for a flight response, indicating that, unconsciously perhaps, the awareness that nature is *not* completely conquered or controlled is still very much alive) is symptomatic of this attitude towards nature. Perhaps his scream therapy can be regarded as an effort to reach out to nature and to atone for his thoughtlessness after all. But overall, the skiing industry as represented in *Turist* exemplifies the experience economy (Pine and Gilmore 1999), fabricating commodified events for customers, so that experiences become commodities (tourists as consumers of 'quality time'). Such fabricated experiences bar us from a genuine nature experience, rather than opening us up to it, so that skiing is not really a 'nature sport' (Krein, 2014, 2015).

Ebba's and Tomas's decision to travel to the Alps is not motivated by a desire to build a rapport with nature, but rather by ego-centric motives. They want to know themselves and each other, unconsciously perhaps, by addressing the question how they will respond to the social, psychic and physical challenges involved in such a pursuit. In other words, although there is continuity between research and skiing (as Sartre indicated), the noema of their quest is not a natural object, but rather their own persona, their own personal Self. Etymologically speaking, the term *persona* means mask, and the avalanche suddenly removes the mask of self-deceit in order to reveal something basic about ourselves (γνῶθι σεαυτόν). Exposure to hazards results in self-exposure. Skiing is an egocentric rather than an eco-centric practice. Besides certain levels of enjoyment (in combination with frustration, irritation, injuries, and the like), it involves test-like experiences (personality tests), triggering self-reflection. From a neuroscientific perspective, the movie emphasises the inadequacy of the neuro-centric conviction that we *are* our brain. *Turist* reveals the extent to which conditioned responses continue to rely on the socio-technological environment as a scaffold. As soon as this world collapses (due to a phantom crisis), research subjects are thrown back on their unconditional responses.

Is skiing an 'urban' outdoors sport (performed on artificial lawns or constructed grounds, deliberately made smoother, squarer, colder, etc.) or rather a 'remote' sport (performed in natural environments, characterised by self-propelled activity and limited infrastructures, taking *what is there* as playing field) as described by Howe (2008)? It is something which hovers in between. In the *Turist* experience, gliding amidst snowy vistas is intimately entangled with technological forms of transport, motorised vehicles, ski-lifts, groomed slopes and artificial snow. Skiing 'artificialises' and urbanises nature (Howe 2008, 3), and although (as Howe phrases it) such semi-urbanised, pseudo-remote activities inevitably 'lose their potential to develop self-understanding and personal growth by offering various kinds of tests' (2), *Turist* indicates that test-like situations may nonetheless unexpectedly resurge.

This leads to a differential diagnostics. On the negative side, there is a tendency in skiing (even in Sartre as a philosopher-skier) to disavow the ecological and technological dimensions of skiing, so that the nature experiences evoked by this practice are much less natural than they purport to be, while environmental awareness is silenced rather than promoted. Skiing as a sportive practice for the affluent masses represents an urbanisation of the sublime, symptomatic for the current era: the anthropocene (Lemmens and Hui 2016). On the positive side, the basic objective of skiing is self-knowledge: putting yourself and others to the test in a carefully designed environment in order to realise to what extent our reflexes and personality traits depend on our supportive world rather than on our brain, followed by therapeutic sessions. *Turist* is basically a 'morality play' (McNamee 2007) and skiing cinema a moral education tool.

Note

1. A fragment of the scene is available on the Internet: https://www.youtube.com/watch?v=unbEUpLkIcI

Acknowledgement

This article is based on a lecture presented and discussed at the 2017 BPSA/EAPS Conference at Radboud University Nijmegen, The Netherlands, April 25 2017.

Disclosure Statement

No potential conflict of interest was reported by the author.

ORCID

Hub Zwart ⓘ http://orcid.org/0000-0001-8846-5213

References

ARISTOTLE. 1980. *Physics (Aristotle 4). The Loeb classical library*. Cambridge: Harvard University Press. London: Heinemann.
ARISTOTLE. 1982. *Art of Rhetoric (Aristotle 22). The Loeb classical library*. Cambridge: Harvard University Press. London: Heinemann.
ATHERTON, JOHN. 2007. Philosophy outdoors: First person physical. In *Philosophy, risk and adventure sports*, edited by Michael McNamee. London: Routledge: 43–55.
BERNARD, CLAUDE. 1865/1966. *Introduction à l'étude de la médecine expérimentale* [Introduction to the study of experimental medicine]. Paris: Garnier-Flammarion.
BREIVIK, GUNNAR. 2010. Being-in-the-void: A Heideggerian analysis of skydiving. *Journal of the Philosophy of Sport* 37 (1): 29–46.
BREIVIK, GUNNAR. 2011. Dangerous play with the elements: Towards a phenomenology of risk sports. *Sport, Ethics and Philosophy* 5 (3): 314–30.
BROCKE, BURKHARD, KLAUS-PETER LESCH, DIANA ARMBRUSTER, DIRK A. MOSER, ANETT MÜLLER, ALEXANDER STROBEL, and CLEMENS KIRSCHBAUM. 2010. Stathmin, a gene regulating neural plasticity, affects fear and anxiety processing in humans. *American Journal of Medical Genetics B: Neuropsychiatric Genetics* 153B (1): 243–51. doi:10.1002/ajmg.b.30989.
CANNON, WALTER. 1920. *Bodily changes in pain, hunger, fear and rage*. New York, NY: Appleton.
CHAPPELET, JEAN-LOUP. 2008. Olympic environmental concerns as a legacy of the winter games. *International Journal of the History of Sport* 25 (14): 1884–1902. doi:10.1080/09523360802438991.
CHIVERS, JOHN. 1994. *Effect of the skiing industry on the environment*. Conventry: School of International Studies and Law.
CONTI, FIORENZO, and SILVANA IRRERA CONTI. 2003. On science and literature: A lesson from the Bernard-Zola case. *Biology in History* 53 (9): 865–869.
CROSSON, SEÁN. 2013. *Sport and film*. London: Routledge.
DOUGHERTY, A.P. 2007. Aesthetic and ethical issues concerning sport in wilder places. In *Philosophy, risk and adventure sports*, edited by Michael McNamee. London: Routledge: 94–105.
FREUD, SIGMUND. 1917/1940. *Vorlesungen zur Einführung in die Psychoanalyse* [Introductory Lectures on Psychoanalysis]. *Gesammelte Werkte XI*. London: Imago.
FREYTAG, GUSTAV. 1863. *Die Technik des Dramas* [Technique of the Drama]. Leipzig: Hirzel.
GEBAUER, GUNTHER. 1993. Sport, theatre, and ritual: Three ways of world-making. *Journal of the Philosophy of Sport* 20: 102–106.
GOTTSCHALL, JONATHAN. 2012. *The storytelling animal: How stories make us human*. New York, NY: Houghton Mifflin Harcourt.
HOWE, LESLIE A. 2008. Remote sport: Risk and self-knowledge in wilder spaces. *Journal of the Philosophy of Sport* 35 (1): 1–16. doi:10.1080/00948705.2008.9714724.
JANICAUD, DOMINIQUE. 2015. *Heidegger in France*. Bloomington: Indiana University Press.
KATWALA, AMIT. 2016. *The Athletic brain: How neuroscience is revolutionising sport and can help you perform better*. New York, NY: Simon and Schuster.
KEENAN, FRANCIS. 1973. The athletic contest as a 'tragic' form of art. In *The philosophy of sport: A collection of original essays*, edited by Robert Osterhoudt. Springfield, MA: Thomas: 309–326.
KREIN, KEVIN. 2014. Nature sports. *Journal of the Philosophy of Sport* 41 (2): 193–208. doi:10.1080/00948705.2013.785417.
KREIN, KEVIN. 2015. Reflections on competition and nature sports. *Sport, Ethics and Philosophy* 9 (3): 271–286.
KRETCHMAR, SCOTT. 1975. From test to contest: An analysis of two kinds of counterpoint in sport. *Journal of the Philosophy of Sport* 2: 23–30.
KRETCHMAR, SCOTT. 2017. Games and fiction: Partners in the evolution of culture. *Sport, Ethics and Philosophy* 11 (1): 12–25. doi:10.1080/17511321.2017.1294554.

LACAN, JACQUES. 1960-1961/2001. *Le Séminaire VIII: Le transfert* [Seminar VIII: Transference]. Paris: Éditions du Seuil.

LACAN, JACQUES. 1962-1963/2004. *Le Séminaire Livre X: L'Angoisse* [Seminar X: Anxiety]. Paris: Éditions du Seuil.

LACAN, JACQUES. 1964/1973. *Le Séminaire XI: Les quatre concepts fondamentaux de la psychanalyse* [Seminar XI: The four fundamental concepts of psychoanalysis]. Paris: Éditions du Seuil.

LACAN, JACQUES. 1971-1972/2011. *Le Séminaire XIX: … Ou pire* [Seminar XIX: …Or Worse]. Paris: Éditions du Seuil.

LEACH, JOHN. 2004. Why people 'freeze' in an emergency: Temporal and cognitive constraints on survival responses. *Aviation, Space, and Environmental Medicine* 75 (6): 539–542.

LEDER, DREW. 1990. *The absent body*. Chicago: The University of Chicago Press.

LEMMENS, PIETER, and YUK HUI. 2016. Apocalypse now! Peter Sloterdijk and Bernard Stiegler on the anthropocene. *Boundary* 2. www.boundary2.org.

LONDON, JACK. 1903/1981. *The call of the Wild/White Fang*. Toronto etc.: Bantam Books.

MARTEL, G., A. NISHI, and G.P. SHUMYATSKY. 2008, September. Stathmin reveals dissociable roles of the basolateral amygdala in parental and social behaviours. *Proceedings of the National academy of Sciences of the United States of America* 105 (38): 14620–14625. doi:10.1073/pnas.0807507105.

MARTIN, ANDY. 2010. Swimming and skiing: Two modes of existential consciousness. *Sport, Ethics and Philosophy* 4 (1): 42–51. doi:10.1080/17511320903264206.

MARTÍNKOVÁ, IRENA. 2011. Anthropos as kinanthropos: Heidegger and Patočka on human movement. *Sport, Ethics and Philosophy* 5 (3): 217–230. doi:10.1080/17511321.2011.602573.

MARTÍNKOVÁ, IRENA, and JIM PARRY. 2016. Heideggerian hermeneutics and its application to sport. *Sport, Ethics and Philosophy* 10 (4): 364–374. doi:10.1080/17511321.2016.1261365.

MCNAMEE, MICHAEL. 2007. *Sports, virtues and vices: Morality plays*. London, New York: Routledge.

MORAN, DERMOT. 2000. *Introduction to phenomenology*. London, New York: Routledge.

ÖSTLUND, RUBEN. 2014. *Force Majeure/Turist*. Beofilm et al.

PAVLOV, IVAN. 1951. Relations between excitation and inhibitions: experimental neurosis in dogs. *Works* III (2): 25–50. Moscow: Progress.

PINE, JOSEPH, and JAMES GILMORE. 1999. *The experience economy*. Boston, MA: Harvard Business School Press.

PITT, REBECCA. 2013. Play and being in Jean-Paul Sartre's being and nothingness. In *The philosophy of play*, edited by Emily Ryall, Wendy Russell and Malcolm MacLean. Abingdon: Routledge: 109–119.

ROELOFS, KARIN. 2012. *Hersenen op de vlucht: het angstige brein in actie* [Brain in flight: The anxious brain in action]. Nijmegen: Radboud University.

ROELOFS, KARIN. 2017. Freeze for action: Neurobiological mechanisms in animal and human freezing. *Philosophical Transactions of the Royal Society B* 372: 20160206. doi:10.1098/rstb.2016.0206.

ROELOFS, KARIN, MURIEL HAGENAARS, and JOHN STINS. 2010. Facing freeze: Social threat induces bodily freeze in humans. *Psychological Science* 21 (11): 1575–1581.

SARTRE, JEAN-PAUL. 1943. *L'être et le néant. Essai d'ontologie phénoménologique* [Being and Nothingness: An Essay on Phenomenological Ontology]. Paris: Gallimard.

SLOTERDIJK, PETER. 2001. *Nicht gerettet. Versuche nach Heidegger*. Frankfurt: Suhrkamp.

SMITH, G.P. 2008. Unacknowledged contributions of Pavlov and Barcroft to Cannon's theory of homeostasis. *Appetite* 51 (3): 428–32. doi:10.1016/j.appet.2008.07.003.

TODES, SAMUEL. 2001. *Body and world*. Cambridge, MA, London: The MIT Press.

VANNATTA, SETH. 2008. A phenomenology of sport: Playing and passive synthesis. *Journal of the Philosophy of Sport* 35 (1): 63–72.

WELTERS, RON. 2016. On ascetic practices and hermeneutical cycles. *Sport, Ethics and Philosophy* 10 (4): 430–443. doi:10.1080/17511321.2016.1201526.

WIPF, SONJA, CHRISTIAEN RIXEN, MARKUS FISCHER, BERNARD SCHMIDT, and VERONIKA STOECKLY. 2005. Effects of ski piste preparation on alpine vegetation. *Journal of Applied Ecology* 42 (2): 306–316.

ZEILER, KRISTIN. 2010. A phenomenological analysis of bodily self-awareness in the experience of pain and pleasure: On dys-appearance and eu-appearance. *Medicine, Health Care and Philosophy* 13: 333–342. doi:10.1007/s11019-010-9237-4.

ZIMMERMANN, ANA, and SORAIA SAURA. 2017. Body, environment and adventure: Experience and spatiality. *Sport, Ethics and Philosophy* 11 (2): 155–168. doi:10.1080/17511321.2016.1210207.

ZOLA, ÉMILE. 1880/1923. *Le roman expérimental*. Paris: Charpentier.

ZWART, HUB. 2008. *Understanding nature: Case studies in comparative epistemology*. Dordrecht: Springer.

ZWART, HUB. 2014. Limitless as a neuro-pharmaceutical experiment and as a Daseinsanalyse: On the use of fiction in preparatory debates on cognitive enhancement. *Medicine, Health Care & Philosophy: A European Journal* 17 (1): 29–38.

ZWART, HUB. 2015. The call from afar: A Heideggerian–Lacanian rereading of Ibsen's *The Lady from the Sea* (published online). *Ibsen Studies* 15 (2): 172–202. doi:10.1080/15021866.2015.1117854.

ZWART, HUB. 2016. Laboratory alice: A Lacanian Rereading of Lewis Carroll?s Alice-Stories as anticipatory reflections on experimental psychology and neuroscience. *American Imago* 73 (3): 275–305.

Intentional and Skillful Neurons

Jens Erling Birch

ABSTRACT
In the mid-1990s, there was a major neuroscientific discovery which might drastically alter sport science in general and philosophy of sport in particular. The discovery of *mirror neurons* by Giacomo Rizzolatti and colleagues in Parma, Italy, is a substantial contribution to understanding brains, movements, and humans. Famous neuroscientist V. S. Ramachandran believes the discovery of mirror neurons 'will do for psychology what DNA did for biology' (http://www.edge.org/3rd_culture/ramachandran/ramachandran_p1.html). Somehow mirror neurons have not received the deserved attention in the philosophy of sport, but perhaps now is the time to reflect on some implications and consequences. The discovery of mirror neurons may increase our insights about our ability to learn, understand, intend, and produce skillful motor actions. In this article I will first examine what mirror neurons are and how they function in monkeys and humans. Second, I will review some objections to the so-called mirror neuron theory of action understanding, and try to reconcile some of these objections. Third, I will inquire into some implications for philosophy, which I believe are also fundamental to several topics in the philosophy of sport. I will then try to relate some of the most interesting aspects of mirror neurons to recent debates in the philosophy of sport. Finally, I will speculate on what further neuroscientific research might teach us about the nature of being a moving subject.

Finding Mirror Neurons, Finding Out How They Work

The discovery of mirror neurons happened almost accidentally. While Rizzolatti and colleagues were doing single-cell recordings on macaque monkeys grasping objects, something extraordinary happened:[1] when an experimenter grasped a cup of coffee, the same neurons fired when the monkey *observed* the experimenter grasping as had fired when the monkey itself had *executed* the act of grasping. Neurons were found in the brain's motor area F5 that responded both when a monkey performed a motor act, and when the monkey observed another monkey or human performing the same act. That was an astonishing incident. It meant action *observation* causes in the observant the automatic activation of the same neural mechanism triggered by action *execution*: the same neurons that are involved in the footballer's execution of the penalty kick might also fire in the brain of the observer (e.g. the goalie) of that kick. Rizzolatti and colleagues established the hypothesis that contrary to the belief that the motor system was a purely executive system in a serially organized brain, there were actually neurons that had both visual *and* motor properties, working in parallel. The now famous mirror neurons allow a direct matching between the visual description of an action and its execution. Mirror neurons match movements we observe to movements we can do, and help us understand the actions of others. This is to say that mirror neurons are active to help us understand what is going to happen next when a teammate or opponent is about to make a pass. Let's make a further inquiry into the properties and function(s) of mirror neurons.

It was long believed that neurons and different brain regions had distinct functional properties. Not so. The posterior motor areas (F1–F5) are heavily connected to the parietal lobe and the cingulate cortex, suggesting that sensory information from the parietal lobe is used to organize and control movement by coding the space around us. The motor system works in parallel with the sensory system so that we are able to differentiate objects and implement movements. When long-term planning and intentions associated with the cingulate cortex kick in, we may begin to talk about *actions*, not mere mechanical movements. Neuroscientific theories claiming that processes in the brain are widely distributed and work in parallel is not new (see e.g. Baars 1997; Changeux 2004; Edelman 1992), so what is? It is the scientific understanding that perceiving and planning could not be done without a moving body in a world. Without referencing the body, it is impossible to apprehend the distance, orientation, and possibilities of objects. What philosopher Evan Thompson (2007, 13–15) calls the enactive approach (see also Ilundáin-Agurruza (2014a) for a nice overview of positions in the so-called program of embodied cognition) is a project trying to 'naturalize' phenomenology by integrating neuroscience, psychology, and philosophy. Such an approach could lead to a Khunian paradigm shift, away from a deconstructive understanding of the body and human behavior. The parallel workings of the brain are crucial to acting appropriately in the world, and how mirror neurons function implies that the motor system is both part of and a cognitive system in itself.

The best way for evolution to make things happen fast and fluid is to give neurons *more* than one property. Rizzolatti and Sinigaglia (2008b, 21) provide an example: picking up an object, say a ball, is a combination of two processes, reaching and grasping. It may seem that reaching precedes grasping, but neuronal recordings show that grasping starts simultaneously as the arm moves to reach. The hand assumes the shape needed to grasp instantly! To grasp something, activation of the primary motor cortex (F1) is required.[2] F1 does not respond to visual stimuli and cannot transform the geometrical properties of an object to make an appropriate grasp. This is done in the F5 (Rizzolatti & Sinigaglia 2008b, 22). F5 does not code individual movements but motor *actions*, which are goal directed and hence intentional. Single-cell recordings show that bending a finger when scratching does not activate the same neurons as bending a finger to grasp (Rizzolatti & Sinigaglia, 2008b, 23). The same mechanical movement has a different *meaning* (they have different (motor) *goals*), and hence different neurons are activated.[3] The mechanical movements in a coin toss and a tennis serve toss might be identical, but since the goals, and hence meaning, are different they have different neural activation. A portion of these neurons in F5 are called *canonical* neurons and respond selectively also to visual objects: an individual neuron fires both when a ring is grasped for (motor property), and when only seen (visual property). The same neuron does not fire when a square is seen, or grasped for (Rizzolatti & Sinigaglia, 2008b, 26–29).

Through what Edelman (1992) calls 'neural Darwinism', or pruning of synaptic connectivity, we learn how different motor responses lead to efficient prehensions (see also Birch 2010). Rizzolatti and Sinigaglia's (2008b, 35, 46–47) interpretation is that F5 contains a vocabulary of motor acts so that we have a repertoire which is at the basis of cognitive functions usually associated with the visual system. Some of the neurons in F5 then discharge both when we observe, and when we do: the visual and motor responses have the same functional meaning because the (motor) goal is identical.

Unlike canonical neurons, the mirror neurons in F5 are not activated when observing objects, but when motor actions involve object interaction (Rizzolatti & Sinigaglia, 2008b, 79–80). Mirror neurons come in classes of specific acts, like grasping, reaching, and holding. There are two major types of mirror neurons: *strictly congruent* and *broadly congruent*. Some mirror neurons show a strict correspondence between the observed and the executed motor act. Others show a correspondence in the goal of the observed and the executed, but not in the precise movements to achieve the goal (Fogassi et al. 2005). If a mirror neuron is strictly congruent, it means the observed action and the executed action have virtually identical neural activation. If a mirror neuron is broadly congruent, then the observed and executed acts have overlapping though not identical neural activity. Most of the mirror neurons (70%) seem to be broadly congruent (Rizzolatti & Sinigaglia, 2008b, 82–84). Finding a strictly congruent mirror neuron is a task for single-cell recording, and in the myriad of neurons very difficult to detect. Finding broadly congruent neurons can be done by techniques like functional magnetic resonance imaging (fMRI), which has been done extensively on human brains.[4] There is a question, of course, as to whether we should apply the term '*mirror* neuron' to a neuron that is not strictly congruent. This is perhaps why neuroscientists have come to speak of a mirror neuron *system* (MNS), which reflects congruent activity in the same brain region when observing and executing. Upon philosophical scrutiny, we might argue that only a system with properties which *appears* to be like mirror neurons has been detected in humans since there has only been one single-cell recording of what is regarded as mirror neurons in humans (we will return to skepticism regarding mirror neurons below).[5] This aside, what are mirror neurons good for? What is their function?

At first glance, we might believe mirror neurons would benefit imitation of other animals' movements, and hence be beneficial to learning (see below). Rizzolatti and Sinigaglia (2008b, 94–97) argue that imitation is involved in the construction of a motor image used in a preparatory stage. The macaque monkey has mirror neurons but does not imitate; only higher primates do. Mirror neurons must hence have an earlier evolutionary origin and another *primary function*, namely *action understanding*: mirror neurons code the goal of actions/motor acts. Since mirror neurons have both visual and motor properties, visual information and motor knowledge can be coordinated. The motor knowledge we possess can be used when observing others. Rizzolatti and Sinigaglia (2008b, 106) claim motor knowledge is 'of fundamental importance for building a basic intentional cognition'. The mirror neuron system provides the mechanism where an animal combines visual inputs with its motor knowledge to differentiate types of actions, and hence understand the actions of another animal. It means that we do not have to reflect explicitly upon another animal's actions, but rather can understand directly and make responses quickly if needed. But what do we know about mirror neurons in humans? Is the MNS responsible for the athletic ability of 'reading the game'? And is the MNS different in humans and in monkeys?[6]

Mirroring in Humans

Electrophysiological techniques like electroencephalography (EEG)[7] have been used on humans to support the claim that there is a MNS in humans as well as in monkeys. In addition, there have been several brain imaging studies (like fMRI) supporting the existence of a MNS in humans, and more is coming all the time (see e.g. Molenberghs et al. 2012). Brain imaging localizes brain areas and circuits involved in the MNS, so that it is possible to find *where* the human MNS is. Cytoarchitectonic—the cellular makeup of a structure—differences in the human and monkey brain make areas functionally and spatially un-identical, but at least brain imaging might enable us to know where to start looking when new technology comes knocking. With fMRI (and similar brain imaging techniques), there can only be evidence of a mirror neuron *system*, and the neurons involved cannot be claimed to be anything more than broadly congruent.[8] We are not even sure what the homolog of macaque F5 is, but Rizzolatti and Sinigaglia (2008b, 121) suggest it is Brodmann's area 44 (BA 44). Others (see e.g. Morin & Grèzes 2008) have suggested BA 6. Empirical evidence shows that compared to monkeys, the human MNS has more cortical space, responds also to non-object-related arm movements, and codes temporal aspects of individual movements (Rizzolatti & Singaglia 2008b, 115–118, 124). This probably means that the human MNS has more fine-grained action understanding, and may even have other functions. What might they be?

Although macaque monkeys do not imitate, humans do. Might the MNS be involved in imitation, and the transmission from observing actions to learning motor and sporting skills (see Hurley & Chater 2005)? Research on mirror neurons is putting its mark on imitation learning for the sport sciences, nicely reviewed by Vogt and Thomaschke (2007). Rizzolatti and Sinigaglia argue that 'the mirror system is involved in the imitation of acts already present in the observer's motor repertoire, suggesting an immediate motor translation of the observed action' (2008b, 143). They imply that mirror neurons are involved in imitation, but *not* in *learning* completely new movement patterns. In addition to the MNS, we must have a control system that inhibits motor movements. If not, we would replicate every movement we observed. So, if we are to learn by imitation, mirror neurons might be a necessary but not sufficient condition (Rizzolatti & Sinigaglia, 2008b, 148–153).

The subject of observational learning is also of interest to sport science, and mirror neurons might play a role here too (see e.g. Lago-Rodríguez et al. 2014; Stefan et al. 2005; Van Gog et al. 2009), although Rizzolatti downplays this part. If the MNS does in fact contribute to observational learning, it would appear to be beneficial whenever we are observing experts on the driving range, swinging a racket, or throwing/catching a ball. Neuroscientific research on mirror neurons might thus change or influence both instruction and learning protocols in sport and elsewhere: knowledge about how much (or how little) practice (5, 10, or 20 repetitions of a movement sequence) it takes to cause synaptic growth (see Kandel 2006) would be immensely beneficial. Rizzolatti does not *deny* that the MNS impacts learning; he is merely claiming that learning is not the primary function of the MNS. The main reason is that mirror neuron activity is increased by already present (motor) knowledge, so mirror neurons do not seem to be involved in learning movements from scratch. Mirror neuron activity might modify or enhance skills and motor knowledge though. Maybe we should say that learning enhances the MNS more than vice versa; learning a wide repertoire of movements seems to increase activity in the MNS. We will see later that an enhanced MNS has the consequence of recognizing subtle differences in actions, so learning motor skills through physical practice and mirror neuron activity is certainly connected. If a goalie is a skilled penalty shooter, then this seems to be beneficial to understand what another penalty shooter is going to do. The possibility of a save might be higher. If it is action understanding and not learning that is the primary function of mirror neurons, we might claim that mirror neurons contribute more to a philosophical than a pedagogical theory. But there are voices arguing that mirror neurons do not play a role in action understanding at all. Let us listen to one of the loudest and sharpest.[9]

Hickok's Critique—and Some Answers

We must be cautious not to be seduced by motor theories, be they of language, perception, or cognition. As philosophers of sport, it is of course easy to be against intellectualism (see e.g. Noë 2005) and embrace theories which emphasize the importance of embodiment and motor action. Perhaps we are too easily led astray by say the philosophy of Merleau-Ponty, the ecological psychology of Gibson, or in this case the mirror neuron theory of action understanding. Philosophers of sport would be good sports if they were the most skeptical of such accounts. We must not forget that there is overwhelming empirical evidence that cognition can be dissociated from body and movement (see e.g. Milner & Goodale 1995). This is perhaps one of the many reasons why most of traditional philosophy is still occupied with theories not involving body or movement. It is a sound endeavor therefore to scrutinize the mirror neuron theory of action understanding; *especially,* if we favor it and feel that a motor theory (of e.g. cognition) just is *it*. In the following then, I will present some serious objections to the mirror neuron theory of action understanding. I will also give some comments on how/whether these objections can be met. This part will also hopefully give us a deeper understanding of the MNS and the theory of action understanding.

Gregory Hickok (2009, 2014) raises some difficulties for the mirror neuron theory of action understanding, all of which stem from skepticism toward any motor/embodied theory (Hickok, 1229). Hickok's problems are based mainly on a critique of inferring mirror neurons from (macaque) monkeys to humans. I will briefly present some general objections before moving to seven particular problems.[10]

Hickok (2009, 1230–31) initially complains that it is not clear what 'action understanding' means in different texts, by the same or different writers, and in different experiments. This is of course a problem, but not particularly for Rizzolatti and other mirror neuron proponents. This is a problem that encounters any enterprise in its infancy. Nor should we forget that philosophers have argued for centuries, and probably always will, about the meaning of concepts like 'action' and 'intention'. In the philosophy of sport, we are still arguing about what skill is, and what kind of knowledge a skill is (see e.g. Birch 2016; Breivik 2014). We should not demand too much of semantics in the neuro-department, but like Hickok we should urge for consistency. In Rizzolatti and Sinigaglia's pivotal book, they take 'action understanding' to mean 'to immediately recognize a specific type of action in the observed "motor events"', 'movements take on meaning for the observer' who '*perceives the meaning* of these "motor events" and *interprets them* in terms of an *intentional act*' (2008b, 97–98). Action understanding must thus be understood as perceiving and interpreting motor events as intentional acts; as recognizing and differentiating between classes of actions by coding goals. The MNS is regarded as a direct neural mechanism doing this so we can use observations to respond in the most appropriate manner, *without* explicit, conceptual, or reflective thought. Rizzolatti and Fogassi (2014) call this 'automatic understanding': an understanding without inference. Embodied theories of cognition which lack conceptual content are what Hickok generally opposes. Hickok does not deny the existence of mirror neurons per se; he denies that their (primary) function is action understanding.

Hickok (2009, 1231–33) first argues though that the function of mirror neurons is not action understanding in monkeys either: there is simply *lack of evidence* for such a claim. If the core claim of the mirror neuron theory is undermined, there really isn't much left. Hickok's hypothesis is that lesions to F5 in monkeys should disrupt action recognition. But this kind of evidence is scarce or variable. It also follows that if one does have action understanding when F5 is impaired, then mirror neurons are not the sole mechanism for action understanding. Hickok argues there are indeed ways to represent and understand actions without the brain areas of the MNS, and that only a small percentage (15%) of the mirror neurons in F5 seem to code the meaning/goal of a motor act. This would also indicate that action understanding is not the only thing the MNS is involved in, but it might still involve the highest amount of mirror neurons for any one particular function. This does not mean that mirror neurons are not involved in action understanding. It is merely to say that there might be different ways to, and other areas involved in, action understanding. This is Hickok's second objection.

Hickok (2009, 1233) argues that the existence of *other mechanisms for action understanding* suggests that it might be difficult to distinguish action understanding from object understanding. Cells in the superior temporal sulcus (STS) do not have motor properties, but have been found to be involved in action understanding. There might be a circuit coming from STS to F5 suggesting that action understanding is achieved primarily through perceptual object recognition and leaving mirror neurons to a mere executive motor command. That there are other areas involved in action understanding is perhaps not the greatest threat to the mirror neuron theory of action understanding. It is commonly accepted that neuronal activity is widely distributed in the brain (see e.g. Baars 1997), that different brain areas work in parallel and loops (see e.g. Edelman 1992, 2006), and that this differs even at an individual level (see e.g. Changeux 2004). The mirror neuron theory simply claims that mirror neurons (perhaps among many) are indeed involved in action understanding, and at some level even primary (for mouth/hand grasping/gripping).

The third problem is that mirror neurons have been found in *other locations than F5*. That there are mirror neurons outside F5 does not in itself weaken the mirror neuron theory of action understanding. It is an empirical question as to whether the functional properties of mirror neurons are also to be found elsewhere. If so, fine. That might just be an indication that several types of cells in the brain have other functions, properties, or connections than scientists originally believed. We are perhaps beginning to see that the brain is not organized neatly, but is complex to the degree neuroscientist Edelman (1992, 29) calls an ever-changing, intricate jungle.

The fourth problem confronts the grounds for *inferring from macaque to human*. Hickok (2009, 1234–1235) argues we have to consider at least three possibilities concerning mirror neurons in monkeys and humans: (1) mirror neurons do not exist in humans, (2) mirror neurons are exactly the same in monkeys and humans, and (3) mirror neurons in humans have evolved to be involved in higher order cognition. The first two possibilities have not been ruled out empirically, and it is therefore a sound scientific attitude to hold one of these possibilities, according to Hickok. After all, the single-cell recordings on humans by Mukamel et al. (2010) do not set out to *falsify* the existence of mirror neurons. This is not to say that the mirror neuron theory of action understanding regarding monkeys is false. But if one adopts the third possibility, which according to Hickok mirror neuron proponents too often do, it is not necessarily true regarding humans.[11] That there is a difference between mirror neurons in humans and monkeys is due to a general difference between these species. Neuroscientists working within evolutionary theory (see e.g. Gazzaniga et al. 2002, ch. 14) agree that the neural networks and pathways in humans are somehow pruned (see Edelman 1992, 2006), epigenetic (see Changeux 2004), and make new connections and synaptic growth at a completely different level than any other primate, including chimpanzees, gorillas, and macaques (see Le Doux 2002, ch. 3–4). This is evident in our relatively long (social) learning phase, compared to other mammals and primates. Even our gene expression changes throughout life to a great degree (Kandel 2006), and this is a necessary requirement for our formidable ability to adapt and learn (by synaptic plasticity), both in daily life and sport. Our human nature is a nature of nurturing, and so is perhaps the MNS (Calvo-Merino et al. 2005, 1246–48; Rizzolatti & Sinigaglia, 2008b, 130). As we have seen, mirror neurons are linked to the F5 in macaque monkeys and it is not clear what the equivalent brain area is in humans. Hickok soundly questions the empirical basis for extrapolating the existence of a MNS to humans. This problem might be resolved in the (nearby) future if/when more single-cell (or similar) recordings are gathered from humans. Until then, we will basically only have brain imaging like fMRI showing (increased) brain activity in brain regions *supposedly* having mirror neurons. Hickok is of course right when he warns to make claims about functional properties across species when location does not hold across species. But that does not mean there isn't a MNS for action understanding in humans.

The fifth and sixth problems raise the question whether *the mirror neuron system can be dissociated from action understanding* in humans. Hickok (2009, 1235–37) argues that the MNS is not a *necessary* requirement for action understanding in humans since other brain areas without mirror neuron properties are also involved in action understanding, and that we are indeed capable of understanding actions we have never performed ourselves (without having motor knowledge of a certain action). The mirror neuron theory implies that if one cannot produce types of actions, one will also have trouble understanding those actions. But this does not seem to be the case since even people born with serious movement deficiencies seem perfectly qualified to understand the actions of others. Hickok also points to the absurdity of mirroring in the literal sense: we would produce the same movement we observe *if* neuronal events where in fact truly mirrored. This would be counterproductive: meeting a basketball lay-up with a lay-up and not a block. At some point then, action observation, understanding, and production must come apart. That would make the mirror neuron system's properties and functions appear meaningless. Although important, this critique is not as devastating to the mirror neuron theory of action understanding as it seems.

That action observation, understanding, and production *can* come apart both intellectually and neurally does not disprove that mirror neurons have both visual and motor properties. It is just to say that many brain areas and circuits work together. This fact is appreciated by Rizzolatti and Sinigaglia (2008a, 2010). Their claim is that mirror neurons might be a platform from which we can have more efficient and fine-grained understandings of actions. Without the MNS, fast and fine-grained understanding of motor goals might be impaired. Rizzolatti and colleagues have never argued that mirror neurons fire identically on a group level. That would indeed be counterproductive. Mirror neurons are, as stated above, more or less congruent. We have activity in the motor cortex just by visualizing movements (and strongly congruent as well, see e.g. Jeannerod & Frak 1999), but the neural activity is not strictly *identical* because at some level we also have inhibition. Interestingly, Mukamel et al's (2010) single-cell recording of mirror neurons in humans found inhibitory activity in mirror neurons.[12] They argue mirror neurons work as a control mechanism to differentiate the actions of others from those of oneself, and as such inhibit unwanted imitation or action. If we didn't have such a mechanism, we could never visualize, fantasize, or imagine without producing movements. Mirror neurons help us both understand the goal of an action, *and* come up with an appropriate response. The final responsive (motor) action is the sum of several events going on, not only the firing of mirror neurons. In this way, action understanding and production both can and cannot come apart. Without the MNS, a goalie might not have the time to react appropriately when say a hockey shot is fired. This is in fact a problem for the serial information processing theorist, like Hickok seems to be. Given the time course of neuronal events, there simply isn't time for the brain to first calculate an outcome, and then come up with responses if a puck or a tennis ball travels fast enough (see Milton et al. 2008, 44–47). The mirror neuron theory might explain how we are able to respond appropriately then, contrary to Hickok's suggestion that mirror neurons would produce identical actions.

Hickok's (2009, 1237–38) seventh problem concerns the *location of F5 in humans*. If F5 is analog to BA 44/6, then damage to this latter region should cause impaired action understanding. But that does not seem to be the case. Since the exact location(s) of the MNS in humans is not established, this is no broadside to the theory and I will leave it at that. We may summarize Hickok's important critique by urging ourselves not to jump to conclusions, especially when empirical evidence is (relatively) sparse, and activity in the MNS *might not* indicate action understanding. Whether we like it or not, we must also recognize the possibility that action understanding, abstract representation, and thought seem to work quite well in humans without motor knowledge, skills, or maybe even a MNS. Neither must we forget that there *is* a huge amount of evidence (see e.g. Gallese et al. 2009, 105; Rizzolatti & Sinigaglia 2010) supporting the mirror neuron theory of action understanding—and more. Mukamel et al's (2010) single-cell recording suggests mirror neurons exist in several human brain areas: hippocampus, parahippocampal gyrus, entorhinal cortex, and supplementary motor area. They suggest mirror neurons are involved in memory functions, inhibition of imitation, self-recognition, and emotional understanding. Let us continue our discussion with the following attitude: suppose the mirror neuron theory of action understanding is true; what then? What would this mean for philosophy in general and philosophy of sport in particular?

Philosophical Implications

There are several important philosophical consequences if we take the mirror neuron theory of action understanding to be true. One implication of mirror neurons is questioning the intuitive view held by many analytical philosophers that the causal chain of intentional action goes from a desire/belief (in the brain) to the arm in a serial order. Single-cell recordings show that in reaching and grasping for food the arm moves without any *declarative* intention, and contractions in the hand may start before movement of the shoulder (Rizzolatti & Sinigaglia, 2008b, 21–25). That is to say, motor actions are intentional in themselves, *and* information processing theories are undermined. This paves way for two consequences I will focus on here:

(1) Motor actions are intentional and thus cognitive. Epistemology cannot stick to the idea that knowing how (motor knowledge) without propositional content is not proper knowledge.
(2) The mirror neuron theory of action understanding supports a motor theory of mind. A consequence for the philosophy of mind should be an increased interest in the moving body, motor action, and sport.

I will briefly elaborate on these issues, and then try to relate them to more specific concerns in the philosophy of sport.

Motor Actions are Cognitive at their Most Fundamental Level

The view that the motor system is simply an executive system without any perceptive or cognitive elements is challenged by the discovery of mirror neurons. Mirror neurons in the motor system are much more complex than classical cognitivism has recognized. Mirror neurons discriminate sensorial 'information' and code it on the basis of potential acts (gripping, reaching, bringing toward). To separate intention from movement is in this light perhaps a mistake. It is quite seldom (if at all) that we merely move our limbs randomly like autumn leaves; instead, we are goal directed. Without diverging into a discussion of the philosophy of action, we might say that we perform intentional actions, not mere mechanical movements.[13] The mirror neuron theory also aligns well with the phenomenological stance for which movement is crucial for cognition. As we have seen, some of the neurons in F5 discharge both when observing and performing motor actions: the motor and visual responses have the same functional meaning/goal. The consequences are crucial for philosophy, sport science, and philosophy of sport: the mirror neuron theory supports the notion of movement as cognition (see Birch et al. in press). The motor cortex of the brain is thus not merely executing movements, but intentional actions. With the discovery of neurons with both visual and motor properties, the distinctions between perception, cognition, and movement are being more than bridged. They are being intertwined, perhaps even brought together as one. Rizzolatti and Sinigaglia claim that motor knowledge is necessary to understand the intention and goal of actions. The whole idea that actions and intentions are solely guided by declarative beliefs or desires is seriously wounded. Perhaps we should abolish the distinction between knowing that and knowing how all together (see e.g. Stanley and Krakauer, 2013). This would truly change epistemology because analytic philosophy has hardly recognized knowing how as proper knowledge at all, sticking instead to propositional knowledge that can be given truth values. Maybe it's not language after all that is primary in everything that the so-called mind does.

A Motor Theory of Mind?

Philosopher Alvin Goldman (2006) argues that mirror neurons might be the fundament for a simulation theory of mind. Goldman has collaborated with Vittorio Gallese, who is one of Rizzolatti's closest research partners. Goldman attacks what he calls a theory–theory of mind. His attack resembles Moe's (2005) critique of classical cognitivism and information processing theory. Goldman argues that understanding low-level emotions (disgust, fear, anger, surprise, sadness, happiness) is not something we do by means of reasoning (through theory, propositional content, or information processing). Instead, evolution has brought forward a faster and more direct way of recognizing emotions, namely simulation. Goldman uses evidence from cognitive neuroscience to argue that if you have not experienced a basic emotion yourself, your recognition of such an emotion is heavily impaired. In contrast to Hickok's arguments above, Goldman refers to a vast number of lesion studies which show that persons with damage to brain areas (like the amygdala) involved in experiencing a type of emotion (like fear) have problems detecting facial expressions of that same type of emotion. This is an analog to the work of Rizzolatti: if you cannot do, you cannot understand—if you cannot experience, you cannot recognize (see also Birch 2009). If we cannot reduce first-person experience to third-person description, this amounts to saying an experience cannot be known in any other way than by being there/doing it.[14] Rizzolatti and Sinigaglia (2008b, 138) insist that mirror neurons give an observer a first-person grasp of the motor goals and intentions of others that we have yet to find elsewhere. This is also to say that we might actually have a neurophysiological mechanism for what is known as phenomenal consciousness in analytic philosophy. In the philosophy of sport, Birch (2009) argues that phenomenal consciousness of how something feels is an essential part of sporting skills.[15] Goldman argues that the neurophysiological mechanism for understanding emotions is the same as understanding bodily actions.[16] This is why Goldman's work is interesting also to the philosophy of sport: understanding emotions and actions is essentially the same. And if so: knowledge of so-called mental states and motor knowledge is essentially also the same. If Goldman is right (and Hickok wrong), it puts motor knowledge on par with what has traditionally been called cognition and this insight may seriously undermine the (Cartesian) divide between body and mind. An empirically supported motor theory of mind has consequences for the philosophy of mind by questioning whether mind, the mental, and cognition are something essentially different from body and motor action. Our body and motor actions then might very well be the hub of our thoughts.

Mirror Neurons in the Philosophy of Sport

Motor theories of mind are not new to the philosophy of sport. What is new is the empirical foundation the discovery of mirror neurons provides (if Hickok's critiques can be met). Neuroscientific explanations of how the brain and the rest of the body work in conjunction with the world should be welcomed by philosophers of sport who have held similar philosophical views. Neuroscientific discoveries have implications ranging from learning protocols (in sport and elsewhere) to epistemological and metaphysical questions concerning the human body. In this article though, I will highlight a discussion recently brought up again by Hopsicker (2009). When Hopsicker extended the contributions made by Breivik (2007, 2014) and Moe (2005) on consciousness and knowledge in skilled motor behavior, he said it is those things we do not declare that deserve attention. As we have seen, mirror neurons function in the interplay between the non-declarative, intentionality, and prior knowledge. In this section, I will try to relate the discussion above on mirror neurons to the contributions by Breivik, Moe, and Hopsicker.

To get a better grasp of intentional movements in sport, Moe first raised critiques against information processing theories held by cognitivists. Moe's arguments came from Dreyfus'

(1992, 2002) anti-representational/anti-rule account and Searle's (1992) neurobiological theory of consciousness. Breivik criticized Dreyfus' view of absorbed coping. Breivik argues that Dreyfus treats the athlete as mostly mindless, and hence underestimates conscious attention in (elite) performance (see also Birch et al. in press). Hopsicker follows Moe and Breivik by analyzing the 'background' and 'attention'. Hopsicker turns to Polanyi to 'examine kinds of knowing and how our intellect operates at the tacit and focal levels during the learning and performance of complex motor activities' (2009, 76). I hope to make a contribution to this discussion by moving into the domain of contemporary neuroscience.

In the information processing theories Moe (2005) criticizes, the motor system is regarded as an executive system without any perceptual or cognitive elements. Since recordings of mirror neurons suggest that there is a *direct* link between seeing and doing, there is no *processing* of information in the way information processing theories describe. Rizzolatti and Sinigaglia (2008b, x- xi, 3, 17–21) argue that we cannot any longer support the view that perception, cognition, and movement are distinct modes. The neuroscience of mirror neurons gives heavy empirical artillery to Moe's (and Ilundàin-Agurruza 2014a, 2014b) arguments against information processing theories and (classical) cognitivism.[17]

In the discussion of mirror neurons above, we have seen that having motor knowledge is what makes the MNS efficient. Having motor knowledge and a mirror neuron system is certainly a great benefit in evolutionary terms. It means one gets a better prediction of what is going to happen, and a better possibility of making an efficient response. The link to sport is evident: having a well-developed MNS and motor knowledge helps us understand what an opponent is going to do: you are better at predicting the trajectory of a ball even if you have seen only a portion of the opponent's motion. We can even unmask tricks and concealed moves, as in a football dribble. As we have already seen, a grasp starts simultaneously as an arm reaches. When observing someone move, the mirror neurons in our brain coding for a goal-directed and *intentional action* start firing at the first minor twitches of the other person. Mirror neurons combined with motor knowledge enable us to understand a motor action at an incredibly early stage (Rizzolatti & Sinigaglia, 2008b, 110–114). Without mirror neurons, the tennis ball served by Djokovic would probably be way behind Federer before his arm began to move. Thanks to the MNS, already in the throw up an expert begins to understand where the ball is coming and start a countermove. Motor knowledge is part of what Searle and Moe call 'the background', which again makes us both understand and produce intentional actions. Rizzolatti and Sinigaglia (2008b, 124–125) also argue that human action understanding based on motor knowledge is done pre-reflectively and non-conceptually. The human MNS might be interpreted as *tacit* knowledge. As we shall see below, there are several reasons to believe that the MNS is trainable and different at individual level. The MNS then plays a role in what Polanyi (1962) called 'personal knowledge'.

Nurturing a Mirror Neuron System

The 'background' and the tacit knowledge discussed by Moe and Hopsicker are the result of experience. They are nurtured rather than the product of nature. When the MNS is considered the result of evolution, it is perhaps easy to think of it as a static system rising from DNA structures which are not trainable.[18] But empirical evidence from sport-related studies actually supports another interpretation. In an fMRI study by Calvo-Merino et al (2005) on expert ballet dancers, experts in capoeira, and non-expert control groups, the following results emerged: the experts had stronger activation than non-experts in brain areas typically associated with the MNS when viewing videos of ballet and capoeira. But not only that: the expert ballet dancers had stronger activation when watching ballet than capoeira, and the capoeira experts had stronger activation when watching capoeira than ballet. Both expert groups had stronger activation than the non-experts when watching videos of their non-expert domain. These findings suggest that a MNS is important for skillful motor behavior, and might be developed through training and experience. Similar results have been found in basketball players (Aglioti et al. 2008), where also decision-making abilities have been linked to the MNS. This kind of research shows how neuroscientific studies on mirror neurons are directly influencing sport science.[19] The trainability of the MNS (e.g. the difference between experts and novices) is also an answer to the problem raised by Hickok against motor knowledge as a necessary requirement for action understanding: action understanding is a matter of degree, and Calvo-Merino shows us that more/better motor knowledge enables a more fine-grained (better) action understanding. Rizzolatti and Sinigaglia (2008b, 136–138) argue that motor knowledge is decisive to *understand* the *meaning* of actions of others. Hence, it is suggested that motor knowledge is cognitive knowledge.[20] For sporting skills, it may also suggest the following: action understanding in team sports like volleyball, ice hockey, and baseball, and individual sports like tennis and boxing (where we respond to the actions of an opponent) is enhanced by the MNS. Our sensitivity to another's motor goals and intentions is better if we have expertise in the specific motor area. For example, experts have stronger neural activity in the MNS and can predict the outcome of motor actions better than novices on the basis of observing only the initial motor action (see Lago-Rodríguez et al. 2014). Mirror neurons are a part of skillful motor behavior, especially in sports where understanding of others' actions is important. As Rizzolatti's philosophical right hand, Sinigaglia (2009, 320) states: the MNS is trainable so that a wide platform of action production enables action understanding. Simply put: the more you can *do*, the more you *know*, making adequate (and creative) responses easier to come up with.[21] This might not come as a surprise to sporting people, but the mechanisms of the MNS give us insight into what goes on in the brain when we do come up with an adequate response. That is after all considerable scientific progress. Motor knowledge involved in action understanding is background knowledge, but it is also tacit knowledge: motor knowledge involved in action production and understanding does not have to be declarative, or rise to the conscious attention Breivik (2007) claims is *also* important in skillful motor behavior. That we do not have direct declarative conscious access to the neurophysiological mechanisms of mirror neurons is not to say that athletes are nonconscious as Dreyfus claims. It is merely to say that consciousness is directly evident in the actions, if we catch Rizzolatti's drift about motor actions being cognitive in themselves. Athletes do not have to make the background (Polanyi's subsidiary awareness) rise to conscious attention (Polanyi's focal awareness—being aware of the content of thought) because they *know* what they can *do*, and in this lies their ability also for action understanding and appropriate (or surprising) responses. The direct matching mechanisms of the MNS enable us explain how we can be intentional, cognitive, and conscious even though knowledge remains tacit and in the background. The obviously cognitive capacity of understanding intentional motor actions seems to be underpinned by trainable motor skills, which are also tacit and in the background.

Let us now sum up what all the hype is about, and what some consequences are (see also Kilner and Lemon 2013). First and foremost, mirror neurons have visual *and* motor properties.

This again implies that both visual and motor responses have the same functional meaning; they have the same goal. In other words: the discharging of neurons is the same in a person hitting a home run, and in you observing that action. This is also to say that the motor system is not solely a final stage in the brain for execution, but a cognitive system in itself. Rizzolatti and Sinigaglia (2008b, 50–51) describe this as 'seeing with the hand'. If we relate this to debates on knowledge, skill, consciousness, and intentionality by Birch, Breivik, Hopsicker, Ilundàin-Agurruza, and Moe, we see that the properties of the MNS provide the possibility of an extended consciousness where the hand or the hammer 'sees the nail' (Hopsicker, 79–80), and gives us a neurophysiological explanation of why we do not have to reflect declaratively when performing a skill. We do not need to have focal awareness or conscious attention on all the subsidiaries/background because there is a direct link between the object (the nail), the grasping of the hand, and the intention of hammering. The discovery that mirror neurons have visual and motor properties undermines the idea of distinct and serial processes in the brain. There is not a perceptual stage, then a cognitive stage, and finally a motor stage. There are parallels, loops, and sometimes perhaps only one thing. Rizzolatti and Sinigaglia's (2008b, x–xi, 3, 17–21) empirically based claim that there is a *direct* link between seeing and doing lends strong support to Moe's (and Ilundàin-Agurruza's) rejection of information processing theories and (classical) cognitivism. The unity of visual and motor functions in mirror neurons secures a fast and fluid understanding and production of intentional skillful actions. When we try to understand how to do, imitate, remember, and reproduce, mirror neurons also seem to play a role. Mirror neurons are an important tacit component of our background knowledge in sport.

So Far and in the Future

Let's say we take Goldman's simulation theory of mind to mean that understanding *emotion* is an understanding of bodily *action*. If read this way, understanding another's mind is understanding another's body, and vice versa. It is a view of an *extended* mind (see also Clark 2008)—extending the mind beyond brain to body and world. We understand both low-level emotions and motor actions through bodily observation. Recognizing and understanding emotions and motor actions seem to presuppose the ability to experience the emotion/action. There is a link between the mirror neuron theory and the popular discussion of phenomenal consciousness (subjective experience; how something *feels*) and psychological consciousness (e.g. cognitive awareness and understanding; what consciousness *does*) (See Birch 2009; Cappuccio 2017; Chalmers 1996). We might wonder if biological creatures like humans could really have the one without the other. This is an important message to take home: if the neat distinction between phenomenal and psychological consciousness is heavily undermined, then research on human cognition should/could not be continued without including the first-person perspective of phenomenality, even when studying tasks like memory and visual attention. The same goes for research on skills and knowledge in sport (see Birch 2009). How something feels (see Nagel 1974) is an integrated part of remembering things, focusing on a task, and making a decision (see also Damasio 1994). What Polanyi and Hopsicker call 'dwelling' (how subsidiaries merge into focal awareness) must be brought into the picture when studying human cognition, skill, and intentionality. To study mere attention (Breivik's 'conscious attention' or Polanyi's focal awareness) then, without what Polanyi and Hopsicker have called tacit knowledge, Birch's usage of phenomenal consciousness, and

Searle's and Moe's 'background', simply does not make much sense. Why? There just wouldn't be any focal/conscious attention without the background/tacit knowledge. There would be no starting point, no idea of what to focus on, no shoulders for the eye's spotlight to rest upon. This philosophical point, brought forward again by Hopsicker, has drastic consequences for research methodology and goals in, for example, (sport) psychology. We might say that the mirror neuron theory of action understanding supports Ilundáin-Agurruza's (2014a) demand for a more phenomenologically oriented methodology concerning consciousness, skill, and intentionality.[22]

If the interpretive and empirical challenges concerning the mirror neuron theory can be met, then it might be true that mirror neurons will eventually do for psychology what DNA has done for biology. The consequences for sport science are perhaps just as enriching. We have seen that the discovery of mirror neurons has implications for imitation, learning, emotions, intentions, and understanding. Moreover, the mirror neuron theory of action understanding presents us with a fundament for a philosophical theory with at least the following suggestions and intriguing ideas suited for further research:[23]

First and foremost, the mirror neuron theory of action understanding provides an empirically based ground for rejecting both a dualistic and an information processing view of the mind, intentionality, and skill. Instead, the mirror neuron theory provides us with a view in which motor action is cognition in itself. This popular view in the philosophy of sport now suddenly has support from the most advanced neuroscience. The mirror neuron theory of action understanding is a motor theory of understanding the intentions of others. It is a theory of both how cognition and the body work, although the distinction between body and cognition might be eliminated, or at least: reconceptualized (see e.g. Ilundáin-Agurruza 2014b). Furthermore, the theory makes us see that cognitive skills and motor skills are at heart (or at neuron if you will) the same. It supports Breivik's claim that consciousness and skill are intertwined, and that skillful motor behavior is not, as Dreyfus argues, mindless (see also Birch et al. in press). Treating non-declarative motor actions as cognitive makes the distinction between knowing that and knowing how blurry. Attempts to reduce knowing how to knowing that (see Stanley & Williamson 2001; Birch 2016; Breivik 2014) are undermined. Finally (and more speculatively), if mirror neurons are important in social cognition, we might even have a neurotheory of ethics (see Rizzolatti & Sinigaglia 2008b, ch. 7). Perhaps we have a neural mechanism underpinning empathy and Levinas' (1961) notion of 'the other'. This has implications for numerous ethical issues in the philosophy of sport: violence, sportsmanship, ethos, and cheating. It is a long leap from the mechanism of mirror neurons to doping behavior, but without recognizing emotions or reactions in the other, it is perhaps difficult to establish a personal morality. Combining for example Jeffrey Fry's (2000, 2003) work on emotions and suffering and Goldman's simulation theory on the one side with Rizzolatti's work on mirror neurons and LeDoux's (2002) work on the amygdala on the other might be a start in such a direction.

The mirror neuron theory has far-reaching consequences worth taking seriously in the philosophy of sport. From the fundamental theory of the body as a direct matching organism and not as a serially organized unit, to understanding and doing intentional motor behavior and sharing emotions on the field, mirror neurons are pertinent for several aspects of sport. With mirror neurons, Polanyi's and Searle's philosophy might also have found a neurophysiological fundament not easily swept under the carpet.

Notes

1. A single-cell recording measures neural events (action potentials) in the brain by inserting electrodes into axons and/or dendrites.
2. Primary motor cortex is often referred to as M1. Rizzolatti and Sinigaglia use F1.
3. Rizzolatti and Fogassi (2014) distinguish between (mechanical) *movements* (the flexion of a finger), *motor act* (movements to achieve a specific goal: flexing a finger to grasp), and *action* (a series of linked motor acts: reaching, grasping, and bringing food to the mouth to eat).
4. fMRI measures changes in metabolism or blood flow in the active brain. With fMRI, imaging is focused on the magnetic properties of haemoglobin. The fMRI detectors measure the ratio of oxygenated to deoxygenated haemoglobin—called the blood oxygenation level-dependent effect (BOLD). For a more extensive treatment of the methods of (cognitive) neuroscience, see Gazzaniga et al. (2002, ch. 4).
5. Mukamel et al. (2010) recorded extracellular activity from 1177 neurons in 21 epileptic patients.
6. As I have already stated, there is the problem of attaching mirror neurons in the strict sense to humans due to lack of single-cell recordings.
7. EEG provides a continuous recording of overall brain activity through electrodes placed on the scalp, which measure large, active populations of neurons producing electric potentials (see Gazzaniga et al. 2002, ch. 4).
8. Brain imaging cannot distinguish between inhibitory and excitatory activity in neurons. This means that we can only see similar activation, but not what kind. Although spatial and temporal resolution in fMRI is increasing all the time, neither localization nor firing rate can be established at the level of identity by brain imaging techniques.
9. Other objections have been raised by De Jaegher and Di Paolo (2007), and Hutto (2008). Sinigaglia (2009, 322–325) has tried to conciliate these objections. See also Csibra (2017).
10. Hickok also objects to the lack of empirical support for a generalization of a mirror neuron system to speech recognition (problem number eight). This critique is mostly connected to theories linking mirror neurons and early language learning. It does not seem to be the most crucial issue for philosophy of sport, and will not be discussed here.
11. Mukamel et al. (2010) argue there is no denying the similarities between monkeys and humans regarding the matching mechanism of mirror neurons.
12. See also Vigneswaran et al. (2013).
13. In discussing fine-grained vs coarse-grained individuation of actions, mirror neurons support a coarse-grained approach; probably more coarse grained than say Davidson's account (see e.g. Davidson 1963).
14. An example: analytic philosophers have tried to resolve Jackson's (1986) 'knowledge argument' by claiming what Mary learns is a knowing how which is not considered knowledge, and hence Mary does not know anything new when seeing colors. This answer is perhaps excluded by the mirror neuron theory.
15. See also Chalmers (1996).
16. The same claim is raised by Rizzolatti and Sinigaglia (2008b, 130, ch. 7). Mukamel et al. (2010) found mirror neurons were active in both facial emotional expressions and hand grasping actions.
17. I state 'classical' in parenthesis because connectionism might also be undermined by these discoveries. Evan Thompson (2007) holds that connectionism is a contemporary neuroscientific information processing theory.
18. Evidence has suggested a mirror mechanism in infants as young as 6 months (Rizzolatti & Sinigaglia, 2008b, 327).
19. An excellent review of neuroscientific research on sporting skills is provided by Yarrow et al (2009). They urge 'neuroscientists to consider how their basic research might help to explain sporting skill' (Yarrow et al. 2009, 585). We are probably only beginning to see the impact neuroscience is going to have on sport science and the philosophy of sport.

20. There is a neuronal link between motor knowledge and motor memory (see Mukamel et al. 2010) in Rizzolatti and Sinigaglia's (2008b, 106–114) theory. For a discussion on memory, knowledge, and skill, see Birch (2011).
21. The neurophysiological explanation is: if you do not have the motor knowledge x, you will not have the neural network z necessary for producing motor action y, so when observing someone capable of y and having x and z, your brain cannot have strict congruent neural activity. You may have broad congruence, but of course the similarity will widen with the difference in x and z, which are (some of) the reasons you cannot do y.
22. Although not treated in this article, Rizzolatti and Sinigaglia's (2008b, ch. 3) philosophical considerations lean on Merleau-Ponty. This link should be most interesting to the philosophy of sport.
23. The mirror neuron theory has philosophical relevance for both body and action. The theory argues for the body as the 'great reason' (Nietzsche 1961, 'Of the Despisers of the Body').

Acknowledgment

I would like to thank two anonymous reviewers for constructive comments and suggestions.

Disclosure Statement

No potential conflict of interest was reported by the author.

References

AGLIOTI, S., P. CESARI, M. ROMANI, and C. URGESI. 2008. Action anticipation and motor resonance in elite basketball players. *Nature Neuroscience* 11 (9): 1109–1116.
BAARS, B.J. 1997. *In the theater of consciousness*. New York, NY: Oxford University Press.
BIRCH, J. 2009. A phenomenal case for sport. *Sport, Ethics and Philosophy*. 3 (1): 30–48.
BIRCH, J. 2010. The inner game of sport: Is everything in the brain? *Sport, Ethics and Philosophy* 4 (3): 284–305.
BIRCH, J. 2011. Skills and knowledge - nothing but memory? *Sport, Ethics and Philosophy*. 5 (4): 362–378.
BIRCH, J. 2016. Skills—do we really know what kind of knowledge they are? *Sport, Ethics and Philosophy* 10 (3): 237–250.
BIRCH, J., G. BREIVIK, and V.F. MOE. IN PRESS. KNOWLEDGE, CONSCIOUSNESS and SPORTING SKILLS. IN *MIT PRESS HANDBOOK OF EMBODIED COGNITION and SPORT PSYCHOLOGY*, EDITED BY M. CAPPUCCIO. CAMBRIDGE MA: MIT PRESS.
BREIVIK, G. 2007. Skillful coping in everyday life and in sport: A critical examination of the views of Heidegger and Dreyfus. *Journal of the Philosophy of Sport* 34 (2): 116–134.
BREIVIK, G. 2014. Sporting knowledge and the problem of knowing how. *Journal of the Philosophy of Sport* 41 (2): 143–162.
CALVO-MERINO, B., D. GLASER, J. GRÉZES, R. PASSINGHAM, and P. HAGGARD. 2005. Action observation and acquired motor skills: An fMRI study with expert dancers. *Cerebral Cortex* 15 (8): 1243–1249.
CAPPUCCIO, M. 2017. Flow, choke, skill. the role of the non-conscious in sport performance. In *Before consciousness. In search of the fundamentals of mind*, edited by Z. Radman. Exeter: Imprint Academic: 246–283.
CHALMERS, D. 1996. *The Conscious Mind*. Oxford: Oxford University Press.
CHANGEUX, J.P. 2004. *The Physiology of Truth*. Cambridge: Harvard University Press.
CLARK, A. 2008. *Supersizing the mind: Embodiment, action, and cognitive extension*. Oxford: Oxford University Press.
CSIBRA, G. 2017. Action Mirroring and Action Understanding: An alternative account. In *Sensorimotor foundations of higher cognition. Attention and performance*, edited by P. Haggard, Y. Rosetti, M. Kawato. Oxford: Oxford University Press: 435–439.
DAMASIO, A. 1994. *Descartes' error: Emotion, reason and the human brain*. New York, NY: Avon Books.

DAVIDSON, D. 1963. Actions, reasons, and causes. *Essays on Actions and Events*. Oxford: Oxford University Press: 3–19.

DE JAEGHER, H., and E. DI PAOLO. 2007. Participatory sense-making. An enactive approach to social cognition. *Phenomenology and the Cognitive Sciences* 6 (4): 485–507.

DREYFUS, H. 1992. *What computers still can't do: A critique of artificial reason*. Rev ed. Cambridge, MA: MIT Press.

DREYFUS, H. 2002. Intelligence without representation—Merleau-Ponty's critique of mental representation. *Phenomenology and the Cognitive Sciences* 1: 367–383.

EDELMAN, G. 1992. *Bright air, brilliant fire*. New York, NY: Basic Books.

EDELMAN, G. 2006. *Second nature*. New Haven, CT: Yale University Press.

FOGASSI, L., P.F. FERRARI, B. GESIERICH, S. ROZZI, F. CHERSI, and G. RIZZOLATTI. 2005. Parietal lobe: From action organization to intention understanding. *Science* 308 (5722): 662–667.

FRY, J. 2000. Coaches' accountability for pain and suffering in the athletic body. *Professional Ethics* 9 (3/4): 9–26.

FRY, J. 2003. On playing with emotion. *Journal of the Philosophy of Sport* 30: 26–36.

GALLESE, V., M. ROCHAT, G. COSSU, and C. SINIGAGLIA. 2009. Motor cognition and its role in the phylogeny and ontogeny of action understanding. *Developmental Psychology* 45 (1): 103–113.

GAZZANIGA, M., R. IVRY, and G. MANGUN. 2002. *Cognitive neuroscience: The biology of the mind*. New York, NY: Norton & Company.

GOLDMAN, A. 2006. *Simulating minds: The philosophy, psychology and neuroscience of mindreading*. Oxford: Oxford University Press.

HICKOK, G. 2009. Eight problems for the mirror neuron theory of action understanding in monkeys and humans. *Journal of Cognitive Neuroscience* 21 (7): 1229–1243.

HICKOK, G. 2014. *The myth of mirror neurons: The real neuroscience of communication and cognition*. New York, NY: W.W. Norton & Company.

HOPSICKER, P. 2009. Polanyi's "from-to" knowing and his contribution to the phenomenology of skilled motor behavior. *Journal of the Philosophy of Sport* 36 (1): 76–87.

HURLEY, S., and N. CHATER. 2005. *Perspectives on imitation: From neuroscience to social science*. Cambridge: MIT Press.

HUTTO, D. 2008. *Folk psychological narratives: The sociocultural basis of understanding reasons*. Cambridge: MIT Press.

ILUNDÁIN-AGURRUZA, J. 2014a. Waking up from the cognitivist dream—The computational view of the mind and high performance. *Sport, Ethics and Philosophy* 8 (4): 343–343.

ILUNDÁIN-AGURRUZA, J. 2014b. Everything mysterious under the moon—Social practices and situated holism. *Sport, Ethics and Philosophy* 8 (4): 503–566.

JACKSON, F. 1986. What mary didn't know. *The Journal of Philosophy* 83: 291–295.

JEANNEROD, M., and V. FRAK. 1999. Mental imaging of motor activity in humans. *Current Opinion in Neurobiology* 9 (6): 735–739.

KANDEL, E. 2006. *In search of memory. The emergence of a new science of mind*. New York, NY: Norton & Company.

KILNER, J., and R. LEMON. 2013. What we know currently about mirror neurons. *Current Biology* 23 (23): R1057–R1062.

LAGO-RODRÍGUEZ, A., B. CHEERAN, G. KOCH, T. HORTOBÁGYI, and M. FERNANDEZ-DEL-OLMO. 2014. The role of mirror neurons in observational motor learning: An integrative review. *European Journal of Human Movement* 32: 82–103.

LE DOUX, J. 2002. *Synaptic Self: How our brains become who we are*. New York, NY: Penguin Books.

LEVINAS, E. 1961. *Totality and infinity—An essay on exteriority*. Pittsburgh, PA: Duquesne University Press.

MILNER, D., and M. GOODALE. 1995. *The visual brain in action*. Oxford: Oxford University Press.

MILTON, J., and A. SOLODKIN, S. SMALL. 2008. Why did Casey strike out? In *Your Brain on Cubs*, edited by D. Gordon. New York: Dana Press: 43-57.

MOE, V.F. 2005. A philosophical critique of classical cognitivism in sport: From information processing to bodily background knowledge. *Journal of the Philosophy of Sport* 32 (2): 155–183.

MOE, V.F. 2007. Understanding the background conditions of skilled movement in sport: A study of searle's 'background capacities'. *Sport, Ethics and Philosophy* 1 (3): 299–324.

MOLENBERGHS, P., R. CUNNINGTON, and J. MATTINGLEY. 2012. Brain regions with mirror properties: A meta-analysis of 125 human fMRI studies. *Neuroscience & Biobehavioral Reviews* 36 (1): 341–349.

MORIN, O., and J. GRÈZES. 2008. What is "mirror" in the premotor cortex? A review. *Neurophysiologie Clinique/Clinical Neurophysiology* 38: 189–195.

MUKAMEL, R., A. EKSTROM, J. KAPLAN, M. IACOBONI, and I. FRIED. 2010. Single-neuron responses in humans during execution and observation of actions. *Current Biology* 20 (8): 750–756.

NAGEL, T. 1974. What is it like to be a bat? *The Philosophical Review* 83: 435–450.

NIETZSCHE, F. 1961. *Thus Spoke Zarathustra*. London: Penguin Books.

NOË, A. 2005. Against intellectualism. *Analysis* 65 (4): 278–290.

POLANYI, M. 1962. *Personal knowledge*. Chicago, IL: The University of Chicago Press.

RIZZOLATTI, G., and L. FOGASSI. 2014. The mirror mechanism: Recent findings and perspectives. *Philosophical Transactions of the Royal Society of London, series B Biological Sciences* 369(1644): 1–11.

RIZZOLATTI, G., and C. SINIGAGLIA. 2008a. Further reflections on how we interpret the actions of others. *Nature* 455: 589.

RIZZOLATTI, G., and C. SINIGAGLIA. 2008b. *Mirrors in the brain*. Oxford: Oxford University Press.

RIZZOLATTI, G., and C. SINIGAGLIA. 2010. The functional role of the parieto-frontal mirror circuit: Interpretations and misinterpretations. *Nature Reviews Neuroscience* 11 (4): 264–274.

SEARLE, J. 1992. *The Rediscovery of the Mind*. Cambridge, MA: MIT Press.

SINIGAGLIA, C. 2009. Mirror in action. *Journal of Consciousness Studies* 16 (6–8): 309–334.

STANLEY, J., and J. KRAKAUER. 2013. Motor skill depends on knowledge of facts. *Frontiers in Human Neuroscience* 7: 1–11.

STANLEY, J., and T. WILLIAMSON. 2001. Knowing how. *The Journal of Philosophy* 98 (8): 411–444.

STEFAN, K., L. COHEN, J. DUQUE, ET AL. 2005. Formation of a motor memory by action observation. *Journal of Neuroscience* 25 (41): 9339–9346.

THOMPSON, E. 2007. *Mind in life*. Cambridge: Harvard University Press.

VAN GOG, T., F. PAAS, N. MARCUS, P. AYRES, and J. SWELLER. 2009. The mirror neuron system and observational learning: Implications for the effectiveness of dynamic visualizations. *Educational Psychology Review* 21 (1): 21–30.

VIGNESWARAN, G., R. PHILIPP, R.N. LEMON, and A. KRASKOV. 2013. M1 corticospinal mirror neurons and their role in movement suppression during action observation. *Current Biology* 23: 236–243.

VOGT, S., and R. THOMASCHKE. 2007. From visuo-motor interactions to imitation learning: Behavioural and brain imaging studies. *Journal of Sports Sciences* 25 (5): 497–517.

YARROW, K., P. BROWN, and J. KRAKAUER. 2009. Inside the brain of an elite athlete: The neural processes that support high achievement in sports. *Nature Reviews Neuroscience* 10 (8): 585–596.

Team Spirit, Team Chemistry, and Neuroethics

Andrew Fiala

ABSTRACT
This paper examines the phenomenon of team spirit from a neurobiological point of view. It argues that ethical judgment should be involved in understanding and evaluating the idea. Adopting a liberal individualist point of view helps us understand the phenomenology of team spirit, while also helping us to articulate a critique of communitarian approaches that celebrate the sort of de-individuation that occurs in team spirit. The paper recognizes further complexity in terms of cross-cultural issues, as well as the tendency to toward reductionism. It concludes by imagining a thought experiment involving chemically enhanced team spirit, which shows how our larger ethical framework helps us evaluate possible responses.

Team chemistry and team spirit are often discussed in speaking about sports and other group activities. These phenomena include trust, subtle communication, shared goals, coordinated activity, and harmonious interaction. One significant question in sports psychology is whether team chemistry or spirit is necessary for winning. A team of star athletes without any team chemistry can defeat a team of amateurs. Losing teams can have a great sense of team spirit. But team spirit/chemistry appears to be part of the experience of joy, grace, and harmony among players. At high levels of competition, team spirit/chemistry may be a component of victorious performance. And human beings enjoy the camaraderie and harmonious action that occur within a team.

A small but growing field of research is looking into the neurological underpinnings of team spirit. To my knowledge, however, there is no other work that has explored ethical issues related to team spirit, team chemistry, and neuroethics.[1] This chapter outlines neuroethical issues in light of growing knowledge about how team spirit functions in brains and bodies. This chapter considers naturalistic explanations of the phenomenon of team spirit, while arguing against reductivism. It argues that ethical considerations are important in understanding these phenomena. It applies this, in conclusion, to the question of the morality of using artificial means to enhance team chemistry.

Preliminary Definitions

The terms 'team spirit' and 'team chemistry' are often used interchangeably. The term 'team spirit' evokes a broader notion of social ontology, directing our attention to the phenomenology of team experience, conceiving its experiential constituents in psychological and social terms as a process involving empathy, de-individuation, group-identification, and socially complex spiritual activity. The phenomenology of team spirit is fascinating from the standpoint of personal psychology and social ontology. We can experience a sort of spiritual transcendence and a loss of individuation in team activity that verges upon the metaphysical or religious (see Nesti 2011). Some may see a spooky, metaphysical mystery here—a kind of Hegelian *Geist* into which the individual is sublated. However, there is nothing spooky about group activity. It is a very natural occurrence witnessed in the behavior of flocks of birds and schools of fish. We might more properly use the term 'team chemistry' to describe the biological underpinning of the ways that groups of individuals coordinate their body–brains to accomplish complex activities. Neuro-chemical analysis may even help to explain the euphoria and sense of belonging that occurs when we experience team spirit.

In common parlance, these terms are often interchangeable. But to be clear, the term 'team spirit' aims to capture the phenomenology of what it is like for an individual to *experience* being a part of a team. Team spirit is what is experienced or witnessed in terms of the collective identity, action, style, mannerisms, behaviors, comportment, etc. that are manifested or exhibited by a team and the way that this is experienced in the life of the team members. Some kind of social ontology is needed to understand the phenomenon. The beginning of a definition of 'team spirit' can be found in Gilbert Ryle's *The Concept of Mind*, where Ryle discusses the category mistake of looking at the individual players of a team in order to see where team spirit is located: team spirit is something more than the individual players—it is a property of the group (Ryle 1949). This does not make team spirit a non-natural thing. Rather, team spirit can be explained in terms of bodies, brains, and world—and by acknowledging that group identity, group intentionality, and group action are more than what is found in the minds of individual members. Team spirit has neuro-chemical constituents; and it can be explained in terms of evolutionary processes. But to explain team spirit in naturalistic terms is not to reduce 'spirit to chemistry,' as it were. We can learn much from attempts to understand the neuro-biological substrate of these phenomena and their evolutionary history. But we should not lose track of the lived experience of human beings who enjoy playing together on teams.

Team spirit is usually viewed as a positive and beneficial thing. It is possible for teams to have poor spirit. Teams can be vicious, mean, and immoral. Teams can exhibit bad spirit or display poor sportsmanship. And a domineering team can subvert individuality in pernicious ways. But my focus here is on team spirit as a good thing: it helps the team succeed and it is experienced by team members in a positive way. A team with good team spirit or chemistry not only functions well in the mechanical sense ('like a well-oiled machine' as the saying goes) but it also fulfills the various ethical, social, and esthetic foci of team activity.

The larger ethical, social, and esthetic frame for evaluating good teams depends upon the cooperative mechanical functioning of the group. We say of a good team that somehow it functions in a way that makes the behavior of the whole more than sum of its parts. That holistic idea takes seriously the idea of group behavior, activity, identity, and even group cognition (see Theiner, Allen, and Goldstone 2010). The idea of group cognition holds that the individual members function as parts of a larger whole and that there is functional cognition and activity in the whole. We see this in the natural world in the behavior of flocks of birds, herds of mammals, and schools of fish. Like other social animals, human beings engage in de-individuated group behavior in a way that is similar to what happens in groups of non-human animals.

When teams fail or fall apart, we say that they lack chemistry, which means that they lack the kind of cohesion and flow that is exemplified by successful teams. When we say that a team has good chemistry, this description is usually morally neutral: the word 'good' is not being used here in a moral sense. 'Good chemistry' is often merely a way of describing efficient and cohesive team behavior. But, as we shall see, moral evaluation is also relevant. It is possible, after all, for a team with good chemistry to do evil things.

At any rate, the positive connotation of team spirit is connected with a positive evaluation of the experience of immersion in a team. The sense of 'group flow' or 'interpersonal flow' (in work by Csíkszentmihályi [1990, 1996] and others) can be used to explain this. A different approach emphasizes 'mindsight,' 'integration,' and 'empathy' under the general rubric of 'interpersonal neurobiology' (Siegel 2010). The individuals on the team join together to work as a team, experiencing this de-individuated cooperation in a positive way. This occurs especially in team sports that require complex coordinated action. It also happens in other coordinated efforts: in the military, in music, theater, dance, and in business. When a coordinated effort is successful, the individuals can feel like a higher power is at work in their communal activity. In music, some say that the 'music plays the band.' In sports, the idea that 'there is no "I" in "team"' reflects a spirit of sacrifice, community, and teamwork that connects with the experience of group cohesion and interpersonal flow. Phenomenologists have described this in terms of 'joint-agency' or 'we-agency' and connected it to studies of empathy and shared emotions (see Pacherie 2014 or Salmela and Nagatsu 2016).

How Does Team Spirit/Chemistry Work?

Research into the neurobiological roots of team spirit is just beginning. One fruitful approach concerns the idea of collective efficacy and related ideas about mirror neurons, entrainment, resonance, and synchronicity (Costa and Garmston 2016; Hart 2008). Other accounts focus on the so-called 'social brain,' the evolution of empathy, and the importance of brain chemicals and the prefrontal lobe. The scientific community dreams of explaining the neurobiological roots of complex social behavior. But as Churchland reminds us, 'neurobiological knowledge has not caught up with our wishes' (Churchland 2011).

The fact that we cannot fully explain how brains and bodies synch up with one another does not mean that this is a spooky or miraculous process. Empathy, cooperation, social integration, mirroring, and synchronizing happen all the time, as does emotional contagion and so-called limbic resonance. Brains coordinate across distance and bodies develop synchronicity. One minor example is both intuitive and provocative: people who sing together in choirs develop common breathing patterns and their heartbeats synchronize (Healy 2013). The integrated coordination of brains and bodies is a fact. Coordination, mirroring, cooperation, and empathy are important for the shared behavior of a social species. Human beings are also able to think and talk about this phenomenon. Coordination is facilitated by language. Linguistic communication is itself a complex form of empathy, integration, and synchronicity. Thus a complete understanding of team spirit may help us understand language, communication, and thought—or vice versa, a better understanding of language and communication can help us understand team spirit.

Consider, for example, our ability to match pitch when singing—and the even more complex ability to harmonize and 'blend' in song. We listen, mirror, modulate, coordinate, and harmonize. Some kind of empathy is essential. Mirroring happens in coordinated gestures and shared dance movements. Obvious cases include contagious yawns or infectious giggling. A behavior in one person elicits a similar behavior in another. At the highest levels of performance, teammates appear to be in synch in a way that indicates non-conscious coordination. This might be explained in terms of mirroring behavior and mirror neurons—or in terms of integration, brain–chemistry coordination, and brain–body–world harmonization.

One fruitful approach to this looks to the neuro-chemistry of brain synchronization and the related experience of flow and team identification. Research suggests a chemical basis for some of this, with oxytocin suggested as one chemical catalyst for team spirit (Churchland 2011; Pepping and Timmermans 2012). Churchland suggests that oxytocin, vasopressin, and the dopamine system are important for empathy, pair-bonding, trusting relationships, and other complex social interactions. Other research shows that teammates share a common endorphin level when working out together—suggesting that this common experience helps to explain team bonding (see work by Sullivan and Rickers [2012]; Sullivan, Rickers, and Gammage [2014]). According to Sullivan's research, rowers on a crew team work harder and in synchrony (and have higher pain thresholds), when there is a common neurophysiological experience. Team bonding experiences, shared activity, common goals, similar pain thresholds, and levels of fitness all seem to have a connection with the development of team chemistry. Team spirit likely also includes less physiological and more psychological components such as a shared language, pledges and promises, and a set of attitudes and intentions oriented toward the team.

The Need for an Ethical Framework

Team spirit is often viewed as a high point of athletic achievement. It is connected to sportsmanship and other athletic virtues. The UCLA basketball coach John Wooden explained team spirit as an important part of his 'Pyramid of Success.' He defines team spirit as 'A genuine consideration for others. An eagerness to sacrifice personal interests of glory for the welfare of all' (Wooden 2005, 69). That's an ethical claim, which says that individuals *ought* to place the good of the group above their own benefit. The claim is not best understood in neurophysiological terms. Rather, the point is moral and psychological, focused on individual attitude, ethical dispositions, and shared moral goals. Phil Jackson, the celebrated NBA coach, saw a higher transcendent achievement in team spirit. Jackson explained in *Sacred Hoops*: 'Basketball is a sport that involves the subtle interweaving of players at full speed to the point where they are thinking and moving as one' (Jackson 2006, 17). As Jackson explains, great teams result when great players connect together through the idea that they are working for something higher than themselves. That spiritual connection is also connected to a sense of modesty, courage, tenacity, and sportsmanship.

While this elevated understanding of the idea of team spirit is inspiring, team spirit can also have negative connotations and impact. Teams can do both good things and evil things. There could be a spirited team of villains. It is possible for there to be 'honor among thieves,' as the saying goes. We might imagine a well-coordinated team of cheaters who experience team spirit and celebrate their illicit exploits as a triumph. Team spirit can also contain a suspiciously anti-individualist message. Some notions of team spirit aspire to subordinate the individual to the team and merge individual identity into the collective. These two problems come together in the idea of 'over-conformity' as a problem: athletes can over-conform to a team ethic that is deviant and pernicious (Hughes and Coakley 1991).

Ethical judgment is required in order to sort out pernicious and beneficial forms of team spirit. Beneath this judgment is the fact that something happens to brains and bodies who work together in teams. In order to sort out good and bad forms of team spirit and team chemistry, it is helpful to understand the neurological mechanisms that allow for team spirit. But we should resist a strict reduction to a neurological or biochemical substrate. The reason for avoiding strict reductionism is found in the psychological and ethical issues mentioned here: team spirit results from intentional states, psychological attitudes, moral ideas, social practices, and linguistic processes. While some ambitious neuroscientist might reduce these psychological, social, and linguistic features of team spirit to a brain processes, we do not want to lose track of these higher order processes. As Glannon reminds us, to attempt to reduce ethics to brain function is 'to conflate normative and descriptive ethics' (Glannon 2011, 114).

More would have to be said to fully flesh out the details of a complete neurobiological analysis of team chemistry. But our goal here is to consider ethical implications. Ethical questions include (1) whether there are ethical dangers that arise directly out of the very concept of team spirit; and (2) whether team spirit should be chemically induced (that is whether direct chemical intervention—as opposed to more traditional methods of creating team spirit—is morally permissible).

Liberal Individualism and the Limits of Team Spirit

Let's begin with the question of the ethical value of team spirit itself. There are risks and benefits to the development of team spirit. These benefits and risks are best understood from within a given normative framework in which harms, dangers, and good consequences are defined.

One important fact about team spirit is that individual identity can be lost in immersion into the collective: the team flows best when the 'I' becomes part of a 'we.' For some, this is a highlight of spiritual development. For others, loss of self and immersion in the team is a negative thing, with deleterious effects. Furthermore, team-centered identities can exacerbate social conflict and create pernicious forms of collectivism, which may hover free of the sorts of ethical constraints normally associated with liberal and libertarian moral and political frameworks. The fact that gangs of villains can exhibit team spirit shows us the danger. And yet, despite the risk of pernicious team spirit, most would admit that team spirit is crucial for successful social activity—on the field of sport, in the concert hall, in family life, and in the boardroom. Of course, to make sense of this claim, we need a definition of 'successful social activity.' And we need an explanation of why some forms of team-centered experience are pernicious. This explanation points us toward the broader question of moral theory.

Thus normative questions quickly appear in any analysis of the phenomenon of team spirit. A research agenda in the study of the neuroethics of team spirit would have to begin by clarifying the normative field, how certain sets of values function, and what counts as 'successful social activity.' The author of this paper agrees in principle with values that are typically associated with liberal individualism as conceived in libertarian and liberal political philosophy. There is no time here to argue in favor of these values. Instead, let me outline the basic standpoint of liberal-individualism. The approach holds individual personhood in high esteem; it views collectivism as dangerous to liberty and individuality. Coordinated social activity—from the liberal point of view—is a second-order good, which ought not eliminate the identity of the individual and which ought not violate individual liberty. Individuals do flourish in community. The experience of team spirit, empathy, social integration, and group flow is rewarding. But communal good should not come at the expense of individual liberty, personal identity, and respect for the autonomy and dignity of persons. The liberal/libertarian tradition warns that collectivism can be ethnocentric and racist, fostering an 'us vs. them' approach to the world. The worst forms of team-identity subordinate individuals to the group. The members of groups are oten viewed merely as member, without concern for their individuality or the intersectional nature of their affiliations and identities.

A neuroethical approach helps explain how brain states and neurochemical transmitters function in the context of group immersion or subordination in group identity. No doubt there are some evolutionary explanations of why individuals experience a kind of euphoria or ecstasy when they join together in group activities. Individual immersion in the herd (and the related neurological processes such as empathy, mirror neurons, synchronization of brain chemicals, etc.) can provide an evolutionary advantage. But this sociobiological explanation does not negate the ethical framework provided by the liberal–libertarian tradition, which emphasizes individualism, autonomy, and separation from the herd. As Carr points out in a discussion of group identity in physical education, 'one man's cooperation may be another man's herd instinct' (Carr 1998, 123). The liberal–libertarian tradition criticizes complete immersion in social groups. The liberal–libertarian tradition reminds us that individuation and autonomy are achievements of the body–brain that ought to be celebrated and defended.

We can locate this normative discussion within the history of ethics, social ontology, and political philosophy in the conflict between Hegelian and Kantian approaches. The ideals of moral and political philosophy that we associate with Kant and the human rights tradition are opposed to the sort of collectivism we find in Hegel and others who assert the primacy of organic holism in social ontology. For that reason, we may want to avoid the very notion of 'team spirit,' since it smacks of a Hegelian social ontology that subordinates the individual to the whole—Hegel, after all, constructed an entire social ontology based upon spirit or *Geist*. Hegel's ontology of spirit can result in drastically anti-individualistic conclusions. Hegel infamously explains that individuals are sacrificed on the slaughter-bench of history in service to the world spirit expressing itself through history. To put it in colloquial terms, the old motto 'There is no "I" in "Team"' aims at a kind of subordination to the whole that liberal individualists should properly reject. A Hegelian social ontology may end up justifying a significant amount of sacrifice on the part of individuals in service to the team. Kantians would protest, emphasizing respect for the dignity and worth of individuals. While not denying the fact that the experience of team spirit can occur, the Kantian would caution us against reducing individuals to tools merely to be used in service of the collective.

And yet there is something inspiring about the notion of team spirit and the sense of collaboration, communication, cooperation, and cohesion that occur in teamwork. One rival framework for thinking about this is the communitarian approach, which runs contrary to liberal individualism. William J. Morgan, has applied a quasi-communitarian analysis to sport and sort of 'unity of purpose' found in cohesive teams. Referencing Hegel and communitarian philosophy, Morgan extols the virtues of a genuine community that is bound by a common purpose, explaining that this offers a kind of 'displacement' of the 'atomistic *I*' in favor of a 'communitarian *we*' (Morgan 2006, 179). Communitarianism emphasizes the immersion of individual identity within the ideals and values of the community. This approach privileges harmony, solidarity, loyalty, and membership over liberty and individuality. Communitarians have long been at odds with liberal individualists in ethics and political philosophy. Here we might cite disputes between communitarians such as Charles Taylor, Michael Walzer, Alasdair MacIntyre, Michael Sandel—on the one hand—and liberals and libertarians such as Robert Nozick, John Rawls, Brian Barry, and Ronald Dworkin (see Ezra [2015] or Kymlicka [2002] for discussion). Communitarians emphasize social cohesion and something like team spirit. Liberals and libertarians are reluctant to affirm the importance of these sorts of social identifications.

The concept of team spirit can even extend to very large systems of group identity. We might consider patriotism in terms of team spirit—and certainly there are times at which entire nations join together behind a sports team (as in the Olympics or the soccer World Cup). What is interesting in this phenomenon is the fact that non-players can partake of team spirit (see Kateb 2000, 920). Communitarians will celebrate these sorts of identifications, seeing the vicarious team spirit that experienced in spectator sports as something to be celebrated. Liberals and libertarians will view all of this with skepticism, fearing pernicious effects of patriotism, jingoism, nationalism, ethnocentric chauvinism, and so on. Liberals and libertarians might also point out the negative side of team spirit as found in the local level—in the behavior of soccer hooligans or in the ways that gangs, teams, and rival groups exclude one another and fight.

Cross-cultural analysis further complicates ethical analysis. Consider a classic example of the cross-cultural problem taken from baseball: the role of *wa* (harmony) in Japanese sports and teams. For Japanese athletes, social harmony, cohesion, and team spirit are primary goods. This creates a problem for American players who have played on Japanese baseball teams. American athletes tend to emphasize individual achievement more than Japanese players do. This difference is cultural, ethical, and political. And these differences lead to a different type of training, a different style of play, and even to different behavior in audiences.[2] Further neuroethical analyses of team spirit and chemistry must attend to this cross-cultural difficulty.

Further analysis must also attend to contemporary critiques of liberal individualism. Some feminists might argue that the question of team spirit and the related problem of the subordination of the individual is a gendered question. It is possible that women and men experience individuation differently and that camaraderie, harmonious interaction, and team spirit are experienced differently by those who inhabit different gendered identities. Similar problems arise when considering the work of critical race theorists, or among critics who argue that liberal individualism is a neocolonial and Eurocentric invention. Given that sport can be analyzed from feminist, race-conscious, and post-colonial perspective, the phenomenon of team spirit is also susceptible to such analysis.

With those substantial caveats in mind, the rest of this essay simply assumes a background of liberal individualism that is grounded in Kantian ethics and human rights discourse. From the liberal individualist vantage point, individuality is a moral achievement. The loss of individuality that occurs in immersion in the collective is morally problematic from this point of view. Moreover, it may be that this loss of individuality is either a form of false-consciousness (do we ever really 'lose' ourselves in this way?) or a substantial devolution from the achievement of individuation that is associated with standard modern European accounts of moral development (such as in Kohlberg). To speak in terms of brains and other naturalistic explanations of consciousness, we might suggest that immersion in the collective reverts us to a more 'primitive' sort of cognition, in which the brain's organic individuation process is subverted or overcome. Notice that the language of the last sentence points toward a normative question: is de-individuation a bad thing, which *subverts* individuality or a good thing, which helps us *overcome* the limitations of the ego?

Given the assumption of liberal individualism, let's stipulate that loss of ego is progressive and valuable only when ego loss is accompanied by positive liberal social values. Thus the loss of individuation that occurs in sleep or death is a total deprivation of value. And there is no positive value in so-called 'neuropathologies of the self,' which occur with damage to the right hemisphere of the brain and apparently especially damage to the medial-frontal and orbitofrontal regions (see Feinberg 2001; Ownsworth 2014). Those neuropathologies include amnesia, anosognosia, somatoparaphrenia, and others. Team spirit is a positive value, when it helps us communicate, cooperate, and collaborate—and when it helps us achieve things that are of value within the liberal individualist framework. We would not want to celebrate team chemistry that helps a gang of thieves to go on a rampage. But we may want to celebrate team spirit among a group of athletes who train together and use their skills to defeat a rival team without subordinating their individuality more than is necessary in order to win. Behind these evaluations is a whole form of life in which sports and athletic competitions are permitted, approved, and celebrated—and in which other forms of group activity are not.

Spirit, Chemistry, and Flow

The ethical analysis discussed above connects to the phenomenology of team 'flow' in which teams of individuals work and 'flow' together. Indeed, the difference between a mere group and a team can be understood in terms of Csikszentmihalyi's concept of flow. We might say that a 'team' flows together, while a mere group simply acts together. 'Flow' is a term that aims to capture the positive sense of absorption in a task that is connected to a feeling of mastery and ease of action. While flow involves de-individuation and absorption in the moment, it is viewed as a psychological (and social and ethical) achievement—as a culmination or consummation.

There is evidence that the experience of flow is connected with stimulation of the prefrontal cortex, which is associated with 'cognition, emotion, maintenance of internal goals, and reward processing' (Yoshida et al. 2014). An analysis that focuses on 'team chemistry' would seek to look below the phenomenon of flow in order to find the neurochemical and brain-based substrate. Once we understand that neurobiological substrate we may be able to manipulate the brain in order to artificially enhance 'team chemistry.' We will turn to ethical questions about this in a moment. But let's note here that there is a difference between flow that occurs as an achievement and flow that is artificially induced. Genuine team spirit and the flow of cooperative group activity are the result of training, team bonding, dedication, and hard work. For it to be viewed as an achievement and consummation, it cannot be the result of a bio-engineered shortcut.

Flow may be tied to success and victory; it may occur in immoral activities as well as high-minded ethical activities. Team flow can be enjoyed for its own sake, even when a team that flows together is defeated. We could even, one supposes, find flow in pernicious activities and when joining together on evil teams. The experience of flow is morally neutral, although it is likely that our moral judgments help to inform and shape the experience. To return to the Japanese baseball example, it is more likely that a Japanese player will experience flow in the context of the *wa* (harmony) that is emphasized on a Japanese team than that an American player will experience flow in that context. Our cultural, moral, psychological, and social presuppositions will inform the experience and our evaluation of it.

At any rate, when harmonious and unified action connects the individuals into a team, there is flow. In those cases, we see the emergence of team spirit or the work of team chemistry. In some recent work on the cognitive processing of jazz musicians, the idea of 'group flow' is explained as flow based on 'relational embeddedness' (Gloor, Oster, and Fischbach 2013). In studies of group dynamics in sports and in music, group flow is described as an emergent property of groups: the team 'gels' or comes together as a unit.

Group flow has been described by Sawyer, who finds it in music, theater, and in sports. Sawyer outlines how group flow occurs when there is communication, concentration, shared goals, a sense of relatedness, a loss of ego, equality of participation, familiarity with teammates, a willingness to accept and move forward, and the risk of failure (Sawyer 2007). Such an account is psychological and spiritual—it is not chemical. The discourse of flow does not usually delve into the chemical and neurological substrate of the experience.

This leads us to the reductionist problem. Is the team dynamic best understood in chemical/biological/neurological terms or is it best described in spiritual terms? This difference in the choice of metaphor has significant implications for understanding teams and group dynamics. The dualistic language of team spirit points toward the existence of some entity that cannot be reduced to the members of the team or to the operations of their bodies and brains. There is a long lineage of thinking about teams in such terms, again with Hegel's social ontology as a prime example. Such an approach is ontologically problematic and ethically risky since it hypothesizes a spooky entity to which the individual parts are to be subordinated. On the other hand, a reductionistic account of team chemistry will focus on biophysical processes while rejecting spiritualistic concepts. A spiritual approach will claim that team activity cannot be explained entirely in neuro-biological terms.

Perhaps the best solution is a pragmatic and dynamic approach to the problem, which adopts a 'both-and' approach to the issue, rather than the 'either-or' of spirit vs. chemistry. We know that when teams form, brain-bodies with an evolutionary heritage of social and cooperative activity find themselves in contexts with deep neuro-biological and ethical/cultural roots. Teams operate as spiritual unities when the members are united by a common goal and set of practices, traditions, and meanings. This can also be described from the standpoint of brains and bodies, in physical terms. This dynamic and pragmatic resolution is ethically preferable, since it reminds us to consider *both* performance-based (chemical-biological-neurological) approaches, which can produce behavioral efficiencies *and* spiritual/psychological approaches, which emphasize virtues and practices understood in ethical terms. At the very least, this reminds us that we ought to include ethical (as well as social, cultural, and psychological) concepts when considering team spirit and the related idea of team chemistry.

Performance Enhancement for Team Spirit?

Now let's put all of this into practice in consideration of a neuroethical thought experiment. What if it was possible to dope the brain somehow in order to induce team spirit? Would that be ethical? We already modify our bodies and brains in various ways in order to induce team spirit. Teams eat common food and drink, adopt common sleep habits, and engage in common activities. One common activity that has obvious impact on brains and bodies is drinking alcohol. An old saying holds that teams that drink together win together. Alcohol has long been a part of team-building and community building rituals in the Western world. Leaving alcohol aside, other team-building activities synchronize brains and bodies in a deliberate fashion: dancing, singing, marching, running—even pre-game chants and huddles. This common and coordinated body–brain activity synchronizes the brains and bodies of individuals in order to create team spirit and/or chemistry. Most people do not have significant ethical objections to these sorts of activities.

It is possible that future biotechnological research will find ways to enhance team spirit and create interpersonal flow and cognitive synchronicity through chemical means. We noted at the outset that dopamine levels synch up among teammates, as well as heart rate and other physiological indicators. We also mentioned that oxytocin, serotonin, and vasopressin may enhance team spirit. If it were possible to use some chemical intervention—a 'team spirit' pill or nasal spray to stimulate team bonding—would it be ethical?

Our thinking about this will parallel our thinking about other forms of cognitive and physical enhancement. There is a large and emerging literature on this topic. But consider an example mentioned by Sandel (2009, 88). He points out that the use of stimulants among baseball players is (or was) common and expected, citing a *Sports Illustrated* article about coordinated use of amphetamine among pitchers and catchers. Other cognitive enhancements could be used to help connect team members. If a pitcher and a catcher are not both on the same wavelength (another interesting metaphor used to describe team spirit/chemistry), they will not play as well together. We could imagine a creative coach or team deciding to take a similar regimen of drugs in order to maximize their coordinated activities.

Our conclusions about this question will be connected to our conclusions about other enhancements. A conservative may state that it is simply wrong to use any chemical enhancement for the purposes of success in sports. The conservative approach would assert that team spirit is only good when it is developed through common effort and good old-fashioned hard work; it is only good when it is developed in concert with associated virtues of loyalty, sacrifice, and so on. A more liberal approach might allow some enhancement technologies, provided that the playing field remained fair and there were no significant side-effects. A more libertarian approach may suggest that teams and individuals should be permitted to experiment with team-enhancing chemicals, so long as no one is coerced.

Much of the contemporary moral literature focused on chemical enhancement has focused *individual* enhancement. But do the same moral frameworks apply in thinking about *team* enhancement? The answer is more complicated. The moral question about chemically enhanced team spirit is exacerbated by the question of whether individuals should *ever* induce the sort of de-individuating experience that happens in team bonding. For the good of the team, should an individual partake of the team enhancement chemical? That question is related to the question of what one should be willing to do for the good of the team.

Should you get a blood transfusion, drink alcohol, get a tattoo, get your hair cut, wear certain clothes, and so on, in pursuit of team cohesion?

Libertarian individualists may conclude that it is morally problematic to deliberately subvert your individuality—regardless of the means. To induce team spirit through chemical means takes this already problematic behavior a step further. A libertarian individualist will also be concerned about the risk for abuse and coercion. What if the team spiked the players' Gatorade with a team spirit drug or sprayed de-individuating chemicals into the locker room air conditioning? A famous science-fiction example of this occurs in Aldous Huxley's *Brave New World*, when soma is served in order to keep the populace in check and to create a sort of phony team spirit and conformity. Non-consensual application of the team spirit drug would be a violation of autonomy. But for a communitarian or collectivist, autonomy is not the only good. Indeed, if it could be established that victory and success—as well as the pleasure of experiencing group flow—were assured, one could imagine some coaches, team-owners, or national team organizations arguing in favor of team-chemistry doping. It is possible to imagine some manipulative coach rationalizing the application of a spirit-enhancing drug by thinking that the team and even the individual who was surreptitiously doped would be thankful for the result. Liberal individualists will view such a rationalization as immoral. It would be an obvious violation of liberty and individual autonomy to dose someone with team spirit soma without their voluntary consent.

But what of voluntary self-administration of team spirit drugs? Would it be moral for an individual to take a drug that subverts his or her own individuality in this way? Let's assume that there are no long-term negative side-effects. Let's also assume that the drug is actually effective. And let's assume that this will help the team to do better in competition. Let's also rule out negative peer pressure and the subtle coercion that can occur from members of the team and its coaches. Would it be acceptable for individual team members to take the drug in this heavily qualified case?

Some libertarian individualists will argue that it is always wrong for an individual to subvert his or her individuality in order to join the collective. Some individualist may be reluctant to affirm any kind of team spirit or collective identity, holding that such collectives are uniformly pernicious—whether organically created or chemically induced. But beyond that, the answer is not so clear. If team spirit is good, if it is experienced as pleasurable for individuals to participate in group flow and team harmony, then it is not obvious that it would be wrong to induce team spirit through artificial means. Individuals enjoy team bonding and synchronizing activities. They eat common food. They sleep in common areas. They adopt common schedules. They don common uniforms. Many organic techniques for creating team spirit are experienced as beneficial and pleasurable. An individualist may find all of this to be foolish and even immoral. But those who value collective action will see it as a normal and beneficial part of team experience. A drug to speed up the process may be viewed as a boon and a benefit that facilitates the potential for peak team performance and the pleasurable experience of group-flow.

Conclusion

This paper discussed neurobiological considerations for thinking about team spirit and team chemistry. It avoided a reductionist approach to the topic. It considered how our normative theories inform out thinking about the neuroethics of team spirit. The framework adopted throughout is sympathetic to the emphasis on individual autonomy found in liberal and libertarian traditions, and is critical of communitarian and collectivist approaches. The paper concluded by considering the question of enhancement technologies which could improve team spirit through creative chemistry. Individualists will be reluctant to affirm such enhancements, especially if they risk being coercively employed. But moral evaluation depends upon a broader consideration of the ethical value of team spirit.

Notes

1. Google Scholar results in no hits for 'team chemistry and neuroethics' and only six tangentially related hits for 'team spirit and neuroethics.' The Philosopher's Index resulted in no hits for 'team chemistry or team spirit and neuroethics.'
2. *The Concept of Wa* on PBS: http://www.pbs.org/pov/kokoyakyu/the-concept-of-war/; and Whiting (2009).

Disclosure statement

No potential conflict of interest was reported by the author.

References

CARR, D. 1998. What moral educational significance has physical education? A question in need of disambiguation. In *Ethics and sport*, edited by M.J. McNamee and S.J. Parry, 119–133. London: Routledge.

CHURCHLAND, P.S. 2011. *Braintrust : What neuroscience tells us about morality*. Princeton, NJ: Princeton University Press.

COSTA, A.L. and R.J. GARMSTON. 2016. *Cognitive coaching: Developing self directed leaders and learners*. 3rd ed. Lanham, MD: Rowman and Littlefield.

CSÍKSZENTMIHÁLYI, M. 1990. *Flow: The psychology of optimal experience*. New York, NY: Harper and Row.

CSÍKSZENTMIHÁLYI, M. 1996. *Finding flow: The psychology of engagement with everyday life*. New York, NY: Basic Books.

EZRA, O. 2015. Distributive justice. In *Bloomsbury companion to political philosophy*, edited by Andrew Fiala, 75–94. London: Bloomsbury Publishing.

FEINBERG, T.E. 2001. *Altered egos: How the brain creates the self*. Oxford: Oxford University Press.

GLANNON, W. 2011. *Brain, body, and mind: Neuroethics with a human face*. Oxford: Oxford University Press.

GLOOR, P.A., D. OSTER, and K. FISCHBACH. 2013. JazzFlow—Analyzing 'Group Flow' among Jazz musicians through 'Honest Signals'. *Künstliche Intelligenz* 27 (1): 37–43.

HART, S. 2008. *Brain, attachment, personality: An introduction to neuroaffective development*. London: Karnac.

HEALY, M. 2013. Researchers explore how we sync together in song. *LA Times*, 8 July 2013. Available at http://articles.latimes.com/2013/jul/08/science/la-sci-cardiac-synchrony-choral-song-20130708 (accessed 22 May 2016).

HUGHES, R., and J. COAKLEY. 1991. Positive deviance among athletes: The implications of overconformity to the sport ethic. *Sociology of Sport Journal* 8 (4): 307–325.

JACKSON, P. 2006. *Sacred hoops: Spiritual lessons of a hardwood warrior*. New York, NY: Hachette Books.

KATEB, G. 2000. Is patriotism a mistake? *Social Research* 67 (4): 901–924.

KYMLICKA, W. 2002. *Contemporary political philosophy: An introduction*. Oxford: Oxford University Press.

MORGAN, W.J. 2006. *Why sports morally matter*. New York, NY: Routledge.

NESTI, M. 2011. Sport psychology and spirit in professional football. In *Theology, Ethics and Transcendence in Sports*, edited by J. Parry, M. Nesti, and N. Watson, 149–162. London: Routledge.

OWNSWORTH, T. 2014. *Self-identity after brain injury*. New York, NY: Psychology Press.

PACHERIE, E. 2014. How does it feel to act together? *Phenomenology and the Cognitive Sciences* 13 (1): 25–46.

PEPPING, G. and E. TIMMERMANS. 2012. Oxytocin and the biopsychology of performance in team sports. *The Scientific World Journal* 2012. doi:http://dx.doi.org/10.1100/2012/567363.

RYLE, G. 1949. *The concept of mind*. Chicago, IL: University of Chicago.

SALMELA, M., and M. NAGATSU. 2016. How does it really feel to act together? Shared emotions and the phenomenology of we-agency. *Phenomenology and the Cognitive Sciences* (March): 1–22. doi:http://dx.doi.org/10.1007/s11097-016-9465-z.

SANDEL, M. 2009. *The case against perfection*. Cambridge, MA: Harvard.

SAWYER, K. 2007. *Group genius: The creative power of collaboration*. New York, NY: Basic Books.

SIEGEL, D.J. 2010. *Mindsight: The new science of personal transformation*. New York, NY: Random House.

SULLIVAN, P. and K. RICKERS. 2012. The effect of behavioral synchrony in groups of teammates and strangers. *International Journal of Sport and Exercise Psychology* 11 (3): 286–291.

SULLIVAN, P.J., K. RICKERS, and K.L. GAMMAGE. 2014. The effect of different phases of synchrony on pain threshold. *Group Dynamics: Theory, Research, and Practice* 18 (2): 122–128.

THEINER, G., C. ALLEN, and R.L. GOLDSTONE. 2010. Recognizing group cognition. *Cognitive Systems Research* 11: 378–395.

WHITING, ROBERT. 2009. *You gotta have wa*. New York, NY: Vintage.

WOODEN, J. WITH J. CARTY. 2005. *Coach Wooden's pyramid of success*. Grand Rapids, MI: Revell books.

YOSHIDA, K., D. SAWAMURA, Y. INAGAKI, K. OGAWA, K. IKOMA, and S. SAKAI. 2014. Brain activity during the flow experience: A functional near-infrared spectroscopy study. *Neuroscience Letters* 573: 30–34.

High-level Enactive and Embodied Cognition in Expert Sport Performance

Kevin Krein and Jesús Ilundáin-Agurruza

ABSTRACT
Mental representation has long been central to standard accounts of action and cognition generally, and in the context of sport. We argue for an enactive and embodied account that rejects the idea that representation is necessary for cognition, and posit instead that cognition arises, or is enacted, in certain types of interactions between organisms and their environment. More specifically, we argue that enactive theories explain some kinds of high-level cognition, those that underlie some of the best performances in sport and similar practices (dance, martial arts), better than representational accounts. Flow and *mushin* (mindfully fluid awareness) are explained enactively to this end. This results in a mutually beneficial analysis where enactivism offers theoretical and practical advantages as an explanation of high performance in sports, while the latter validates enactivism.

1. Introduction

Legendary and magnificent tennis rivals Chris Evert and Martina Navratilova—both with 18 Grand Slams singles finals—played against each other in 80 matches. Their games were unerringly highly contested. Take for example their 1918 Australian Open final, which went 6–7, 6–4, 7–5 to Navratilova. Developing a philosophical account for how their skills and cognition enabled them to play their consummately skilled duel of backhands and forehands, finely tuned topspin, angling of serves, gently dropped shots, and forceful smashes, is a formidable challenge.

Theories of mind have generally held that cognition depends on meaningful mental states occurring in systems that construct representations of their environments. According to standard accounts of mental representation then, for Evert and Navratilova to perform, they had to be able to internally represent their own position, and that of racquet, court, ball, and opponent. Their senses informed them of ball and adversary location within the court. Then, their brain processed this information, inferred likely trajectories and displacements for both, and directed their body to move so as to best hit the ball. This way of thinking about cognition has dominated philosophy of mind and cognitive science for at least the past century. It has standardly been thought that cognitive systems, the kind of beings that exhibit intelligent behavior as a result of mental processes, must be representational systems—they must be able to internally represent the objects that their mental states are about.

Enactivism is one of many approaches that emphasize the constitutive role of embodiment in cognition. Rather than tying cognition to representational content, enactive approaches see cognition as arising from interactions between organisms and their environments. According to enactive approaches, when organisms adapt and respond to environmental changes in ways that increase their success, cognition is found in what the organisms *do* in relation to their environment. Cognition is thus enacted.

For those who hold that representation is necessary for cognition, organisms must be able to represent aspects of their environment in order for cognition to take place. On the other hand, because enactivists maintain that at its basic level cognition is a matter of interacting with the environment it does not automatically follow from the fact that an organism lacks the neural hardware needed for representation that the organism cannot engage in cognition. Thus, enactivists argue that relatively simple organisms may be cognitive systems. To be clear, enactivists do not claim that human activities never involve mental representations. Instead, it is generally held that while cognition is enacted in basic minds, most of the sophisticated activities that we consider uniquely human do require representations. Language use, detailed planning, and the development of technology fall into these groups.

We will argue that enactive theories best explain some kinds of high-level cognition, indeed the type of cognition that underlies highly skilled performances in sport and similar practices (dance, martial arts). In particular, the spotlight will be on flow and *mushin* (mindfully fluid awareness) states. These are very sophisticated cases of cognition during performance that involve focus, awareness, creativity, and intention but, as we will contend, do not necessarily involve representation. Of all putative candidates for activities that are conducive to these states, sports and martial arts are the most iconic and among the clearest. Consider a snowboarder dropping on a steep big mountain line on large-scale mountain face. Ideally, she will find herself experiencing flow states when her ability to carve turns matches the difficulty of the terrain and conditions, or alternatively, when the terrain brings out her best riding.

While our main interest is the examination of high-level cognition in sporting contexts, it is important to scout more general territory in the philosophy of mind for several reasons. This will lay the requisite groundwork to understand the relevance of the enactive view for sport philosophy. It will also help to re-conceptualize skilled performance, and offer points of contact with complementary views from other pertinent disciplines (e.g. psychology). Additionally, this analysis is mutually advantageous for sport (and its philosophical examination) and enactivism. Sport validates enactivism and extends its range of application beyond the basic minds that enactivist scholars often discuss; enactivism, brings explanatory transparency when it is applied to highly skilled sporting action.

We begin with a short summary of what enactivism is. Then, we turn to the issues that have beset representational theories and explain some of the advantages of understanding cognition as enactive and embodied. Then, we return to a discussion of sport and the body-minded[1] states associated with high performance, particularly flow and *mushin* (mindfully fluid awareness), and argue that such states are best explained by enactivist approaches. Finally, we conclude by discussing the advantages that an enactive account and sport offer to each other.

2. What is Enactivism? Representational and Enactively Embodied Theories of Mind

After a brief discussion of embodied cognition to contextualize key issues, in this section we turn our attention to some of the problems with representational approaches that have motivated the move toward enactivist approaches. Following this, the radically enactive approach of Dan Hutto and Erik Myin (2013) comes to the fore.

2.1. Embodied Cognition—Overview

The label of 'embodied cognition' is misleading insofar as it does not designate a unified position but rather incorporates a variety of promising yet contentious approaches. Accordingly, its ranks are constituted by embodied, enactivist, externalist (which include extended, embedded, and extensive factions), and situated views. Some assess embodied cognition as an energetic but somewhat conceptually confused program (Krieger 2014), while others find the term 'embodied cognition' to be but a lexical band aid that redundantly speaks of processes already captured adequately by theories concerned with qualitatively felt moving bodies (Sheets-Johnstone 2009, 2011). In short, the embodied label covers a promising but motley array of stances of varying strengths and weaknesses. While some embodiment approaches may have to be abandoned, disagreement should not lead us to conclude that either the general direction is mistaken or that all embodied approaches are flawed. Leaving polemics aside, the key underlying tenet—theoretically, phenomenologically, and empirically supported—is that, contrary to still prevalent views in cognitive science or functionalist stances in philosophy of mind, embodiment is fundamental to cognition and to how the world is *actually* experienced.

We are concerned with a stronger claim that some of those working in philosophy of mind and cognitive science, who share allegiance to the embodied camp, have begun to explore: the possibility of explaining cognition without reference to mental representation. Enactivists argue that cognition is not fundamentally based on contentful internal states, and rather than minds being located in and limited to brains, our cognitive states are 'found' in the continued activities in which we engage. Briefly, enactivism takes action to be constitutive of cognition and perception. Within the enactive cohort are found a number of approaches, such as John O'Regan and Alva Noë's sensorimotor enactivism (2001), which Noë (2004) developed more fully subsequently. In this case, perception is about interacting with the environment. Evan Thompson's (2007) autopoietic enactivism centers on how self-organizing living organisms adapt to their environments in dynamically complex ways. Here, we focus on Dan Hutto and Eric Myin's radical enactive cognition (REC). Hutto and Myin's approach differs from most others in that they reject the idea of locating cognition in the brain and instead see it arising as part of a relation between organisms and environments, which they categorize as 'extensive'. Although we will not argue the point here, we see Hutto and Myin's REC as providing more explanatory power than alternative theories of embodied cognition, aligning with a holistic stance that avoids dualistic terminology and conceptualization (Ilundáin-Agurruza 2016). We *will* argue in particular that the skillful and refined activities found in sport are best explained by REC. If we are correct, much of what we conclude will apply to other approaches to embodied cognition as well.

2.2. Mental Representation

Given that enactive approaches to cognition, including REC, arose against a backdrop of representational theories of mind, it will be helpful to provide an overview of representational theories and then explain how enactive theories compare. We begin by considering why it has seemed that any explanation of mind must include the assumption that representation is a necessary condition for cognition.

First, even if representation can involve sense perception of present objects, the role of mental representation is clearer when it involves concepts of objects (or events) not immediately perceived. Thinking about Mohammed Ali fighting Joe Frazier requires being able to represent Mohammed Ali and Joe Frazier in order to have a thought about their fight. To explain this we may posit the ability to represent propositional content in some way. Another clear case is imagination, e.g. picturing a conversation (or argument) between Socrates and Mohammed Ali in Hades. Imagination differs from other cognitive processes that may involve perception, judgment, and reasoning in that we regularly entertain thoughts about persons or objects that are not present, and may never have been. This seems to require even more complex mental representations.[2] Consider organisms that cannot represent objects, particularly those not directly present to their senses, for example clams, snails, or sardines. These may respond to immediate stimuli, but many people, definitely those with theoretical allegiance to computational accounts of the mind, would not think that they are engaged in cognition proper. Intelligent behavior seems to depend on the ability to respond to hypotheticals, and this seems to require at least a basic ability to represent features that are not currently present.

A second, perhaps even more important, motivation has been the expectation that a theory of mind should be able to explain how it is that thoughts cause other thoughts and behaviors.[3] If it is the case that we can explain the causal links between mental states, these mental states must be identical to, reduce to, supervene on, or stand in a causal relation to, states of some sort in our brains. Nonphysical entities do not have causal powers. So, unless mental states are somehow instantiated in brain states, it seems unlikely that they can play the required causal role. And, whatever that relationship is, it must allow formulating and maintaining representations of the objects of thoughts.

Without internal representations, it is difficult to see how thoughts could cause other thoughts, how they could bring about behavior, indeed how they could play any causal role at all. Without bringing representations into the picture, it seems that we cannot explain how sportspersons manage to carry out their performances. How else than by appealing to some sort of mental representation of a football pitch, ball, and opponents could Lionel Messi be able to successfully feint and score? How could a climber scale up a boulder if not by representing hand and footholds as well as her own limb and body positions? Or, how would we account for two *kendō* fencers' complex and complexly regulated movements during a bout at the *dōjō* (道場)?

Considering examples such as those above, it seems obvious that the way to understand the mechanisms underlying cognition is to understand mental representation. Given that representation was taken to be necessary for cognition, the generally held assumption was that the line between organisms that are capable of cognition and organisms that are not lies somewhere near the ability to represent features of the organism's environment. Thus, rather than asking whether or not cognition is possible without mental representations, most work in philosophy of mind has assumed that cognition requires representation and instead tried to better understand the nature of representational states.

Nonetheless, there are longstanding concerns with the representational approach. While the idea of explicit internal representation fits some activities well—specifically those in which humans formulate thoughts in language—it is not at all apparent that other forms of human activity or intelligent animal behavior are guided by similar internal representations.

Along these lines, Dreyfus (1972) pointed out that much of human behavior, at least from a phenomenological perspective, does not involve conscious representation. Alternatively, analytic philosophers of mind like Daniel Dennett (1987)—especially in his more instrumentalist moods—questioned whether folk-psychological explanations of human behavior actually correspond to, or map onto, internal brain states. Dennett pointed out that assuming a system is rational and attempting to fulfill its needs or desires according to beliefs can generate such explanations from the outside. According to Dennett, this is how we explain and predict the behavior of many intelligent systems, whether or not the beliefs and desires are represented internally or not.

Third, despite all of the attention it has garnered, there is little agreement about what it is for one thing to represent another. There is, for example, an issue concerning the level at which such representations should be dealt with. Very generally, a representation can be understood as anything that stands-in for something else (Shapiro 2011). A map, for example, stands in for the terrain it represents. Philosophers of mind of a naturalist bent are particularly interested in neural structures that stand-in for the things that one's mental states are about. Consider, for example, the Evert and Navratilova again. Supposedly, they have neurologically realized internal representations of the court and the other player. These neural structures stand-in as a kind of internal map. Philosophers ask *how* different structures in the brain come to meaningfully represent features of the environment.[4]

In a general sense, philosophers tend to agree that if a particular thing represents another then covariance is a minimal expectation. When a change happens in the environment and an organism perceives the change, the expectation is that an internal state changes in the organism. But covariance is not all that is needed. If the fence in the backyard sways when the wind blows, the swaying co-varies with the intensity of wind from the north. But clearly the fence does not represent the speed of the wind in the sense that is needed here. Representations have content, and as Hutto and Myin Argue:

> […] anything that deserves to be called content has special properties—e.g. truth, reference, implication—that make it logically distinct from, and not reducible to, mere covariance relations holding between states of affairs. Though covariance is surely scientifically respectable, it isn't able to do the required work of explaining content. (Hutto and Myin 2013, 67)

Much of the philosophical work on representation has been devoted to closing this gap between covariance and contentful representation. Hutto and Myin (2013) argue that it seems implausible that a truly naturalized theory that explains the step from non-representing systems to representing systems is available. When trying to close this gap, or to explain what it is for one thing to represent another, the advantages of enactivism become clearer. Besides the obvious parsimony, the enactive account is expedient not only in terms of philosophical psychology but also in the context of sport and skilled performance. Before examining these benefits, however, one more account and defense of mental representation needs to be considered.

Ruth Millikan's (1984) teleosemantics is perhaps the best candidate for explaining what is needed for something to function as a representation in a basic organism. Millikan's central idea is that, in addition to simple covariance, a structure or device within a system represents a feature of the environment just in case the system uses that device or structure because it has the proper function of representing said (type of) feature. Proper function is explained in terms of natural selection. A structure or device that gives an organism the capacity for fitness in an evolutionary sense because it co-varies with a feature of the environment has the proper function of representing that feature or type of feature. Millikan uses the example of bees that find nectar at a particular place, then return to the hive and dance in a way that leads other bees to travel to the place in which nectar was spotted. The bee dance is a representation because its movements co-vary with the direction and location of nectar *and* because it exists to perform the function of getting other bees to travel to location where there is nectar (and this is its proper function).

Hutto and Myin (2013) argue that although the intent of Millikan's teleosemantics is to naturalize meaning by contextualizing covariance in the process of natural selection, there is still a gap between successfully functioning as a system and actually representing. According to them even if we adopt Millikan's approach, '… we can't think of cognitive agents as content-using, content consuming, or content interpreting systems, for to do so suggests that there is a preexisting content to be dealt with'(Hutto and Myin 2013, 76). The point is that even if we can attribute content to the states of organisms according to how those states contribute to the success of the organism within the context of evolutionary theory, the content of those attributions is not in the features of the world with which the states of the organism co-vary. Further, there is nothing in the systems themselves that could create such content. In other words, content cannot be accounted for either through environment or through the organism.

Nonetheless, Hutto and Myin's point is that it does not follow from this that the behavior is not cognitive. They maintain that the directness found in the types of systems described in Millikan's account does not need to be tied to representational content. As a result, they advocate replacing Millikan's teleosemantics with teleosemiotics:

> In short, it is becoming clear to many in the field that purely biologically based accounts lack the right resources for naturalizing mental states with purely semantic properties, such as truth and reference. If we reject teleosemantics in favor of teleosemiotics, we can borrow what is best from the former as well as accepting covariance accounts of information in order to provide a content-free naturalistic account of determinate intentional directedness that organisms exhibit toward their environment. (Hutto and Myin 2013, 81)

Among those who advocate teleosemiotics is Rowlands (1997), for example. This refined and sophisticated dance around representation is crucial if we are to understand how organisms interact successfully with the myriad changing situations they encounter. After all, whereas representation may be expeditious in one context, it may prove deleterious in another. Carving out the space to allow contentless cognition is imperative.

Hutto and Myin maintain that:

> It is possible to explain a creature's capacity to perceive, keep track of, and act appropriately with respect to some object or property without positing internal structures that function to represent, refer to, or stand for the object or property in question. Our basic ways of responding to worldly offerings are not semantically contentful. (2013, 82)

If Hutto and Myin are correct, then some of humans' most unique endeavors involve cognition without content.

2.3. *Human Cognition and Basic Minds*

The above explains why much of the work in embodied cognition has focused on basic minds. If we understand how organisms with basic minds enact cognition, the idea is that this gets closer to a complete picture of mental activities of both enculturated humans and simpler organisms. But, a lot needs to be filled in. Particularly, if the goal is to account for a well-placed arrow in the bull's eye such that the elucidation bypasses representation in a philosophically rigorous and tenable way. The model for basic minds, to use Hutto and Myin's terminology (2013), behind the enactive program and on which human cognition is mapped on, is that of bees, sticklebacks, and simpler organisms. Even if Hutto and Myin clearly intend to apply this model to humans—with all respect due to bees' cognitive capacities—extending the model to sophisticated human cognition is challenging.

Some of the cognitive states of humans clearly involve representational content. One benefit of the enactivist approach is that it may help to show how representational states might arise from the evolutionary development of basic minds. If one takes the position that representational content is required for cognition, that only behavior based on such content is intelligent, and that such behavior is radically different from all other behavior, then it becomes very difficult, and perhaps impossible, to explain how this radically different type of state arises.

If we reject the idea that all cognition requires contentful mental representations then we open the door to explain how human capacities can develop from intelligent behavior in more basic minds. Rather than there being a wide gap between high-level human behavior and all other behaviors of both humans and other animals, we can attempt to explain how human behaviors that include representational content are built on other cognitive behaviors. And, it becomes less puzzling how the sophisticated behavior of non-human animals can be considered intelligent irrespective of whether it depends on representational content or not. Understanding cognition in systems that clearly do not form internal representations provides the key to understanding cognition in general. If human cognition is scaffolded on the cognition proper of basic minds, then understanding the latter is important.

Of course, this does not answer the question of how contentful human representations arise. This is not the place to attempt to answer this question. We will say, however, that it seems likely that meaningful, full-fledged contentful states only arise in the context of complex communication through language.[5] On the one hand accepting that intelligent cognition does not rely on such representational states begins to sow the ground for how such linguistic capacities might arise. On the other hand, denying that intelligent behavior is possible without representations presents a much wider gap to be bridged.

The resulting picture is one in which human beings share many cognitive, but non-representationally based capacities with other organisms. But, as a result of interacting with other humans and using language, humans also have capacities that allow them to accomplish many things that could not even be understood without complex networks of contentful representations. Our high-level functioning then, our most sophisticated behavior, *seems* to be based on our representational capacity.

3. What Does This Have To Do with Sport?

This section discusses how the enactive approach connects with sport and related practices, first considering it in relation to high performance states, and then detailing how such states can be taken to be enactive.

3.1. High- Level states—Flow & Mushin

Flow and *mushin* both refer to bodyminded states experienced during heightened performance. These are amenable to being experienced across a broad variety of practices. As Csikszentmihalyi (1990) details, flow states arise in contexts that range from the arts and leisure pursuits to work environments. He identified nine characteristics that included extremely focused attention, merging of action and awareness, and elastic time dynamics (time slowing or speeding up), which arise when performers find a balance between their skill level and the challenge they face. *Mushin* (無心), usually translated as 'empty mind' or 'no mind,' is better conceptualized as a mindfully fluid awareness—with the emphasis on being fully attentive *without* either getting caught up or becoming distracted. While such states are commonly referenced in East Asian culture, e.g. in China these are known as *wushin*, they are particularly nurtured in Japanese traditional practices of self-cultivation, *dō* (道), such as in *kendō* (剣道), the art of the sword.

Nature and alternative sport athletes often claim that flow is the reason they pursue their sports.[6] Consider the following explanation from professional snowboarder Jonaven Moore in response to a question about what motivates him,

> It is this feeling that I get surfing, big mountain [riding] is for sure the same for me. Your brain shuts down and you live in these moments, and its something that's really hard to explain, but it is the best feeling ever. (Pauporte, Pauporte and Pauporte 2007, 1:20).

Clearly, Moore's brain does not literally shut down, even if certain areas of his prefrontal cortex are deactivated (see Kotler 2014, 49). Rather, the shutting down is his way of accounting for the hard to explain sensation of 'living in the moment' characteristic of flow states.[7]

While traditional sport athletes also experience flow, it is most often secondary to the competitive aspect of their sports. In the context of a ski or snowboard slalom competition, racers focus not on their experience but on finishing first—on getting down the course as fast as possible without missing a gate.[8] Putatively, this can compromise flow experiences because seeking top performance often involves exceeding personal capacities, which in turn brings anxiety and often curtails flow.

In contrast, consider a master of *kyūdō* (弓道), the way of archery, in one of its contemplative schools. Her goal is not to hit the target per se but to release the arrow 'naturally,' spontaneously, and in the right state of mindful focus. That is, what matters is less the marksmanship than releasing the arrow properly. Put differently, hitting the target is but evidence of right shooting. Moreover, shooting is but a means to a holistic and integrative development of skills and moral/spiritual development proper of master performers (Ilundáin-Agurruza 2016). Of the eight different stages in *kyūdō* archery, two are foremost and predicated on the emptying of bodymind and ego in a state of *mushin* the release, *hanare*.[9] (離れ), is the consummation of shooting. But what allows judging whether the release was good or bad is not hitting the target, but *zanshin* (残心). *Zanshin* refers to a remaining attitude or form evidenced by the posture and demeanor of the archer (Acker 1998). When marksmanship prevails, however, as in some schools with sportive tendencies, this lingering spirit or form is compromised.

Flow and *mushin* states are functionally analogous in that typically these accompany higher performances. They also parallel each other formally in so far as competitive pressures may hinder their occurrence. But, they are not identical; they differ phenomenologically and procedurally (Krein and Ilundáin-Agurruza 2014). Moreover, and besides different kinesthetic dynamics, flow states are usually accidental and not expressly sought in competitive western sports (although this is starting to change, see Kotler 2014), while in Japan *mushin* states are cherished and fostered. In fact, the goal of *dō* is to precisely instill such states, which are seen as outcomes of expertise and moral rectitude. Flow in western sports has traditionally been an ancillary by-product of high-performance, whereas in East Asian martial arts it is the very path to high performance. Moreover, even when engaged in activities premised on risk and high-speed reaction times, such as risk sports like wingsuit flying and traditional *kenjutsu* (剣術), swordsmanship, differences arise at the level of enculturation due to *mushin*'s explicit Buddhist framework (Ilundáin-Agurruza et al. 2017).

In spite of these differences, the shared underlying functional parity between the two kinds of states is important from a cognitive standpoint: to operate at the highest level of skill, they both rely on contentless cognitive states. This is true even of cases of overt and 'thickly' enculturated performance in sports and martial arts, or so we argue presently. The objective now is to explore the way enactivism relates to these states.

3.2. Are These States Enactive?

The critical move is to show that high-level cognitive states are possible without representation. Specifically, the case needs to be made that mental representation, as discussed above, is not necessary in highly skilled sporting performances. In fact, we contend that such level or performance thrives in the absence of representations. Because others have detailed the problems that burden classical cognitivism and other representational-heavy accounts (Moe 2007) and even phenomenological ones such as Dreyfus' (Breivik 2013; Eriksen 2010), we obviate these challenges to pay particular attention to how the enactive account operates in sporting and alike contexts.

On a philosophically broad construal, mental representation concerns objects (most commonly thought of as brain states) that have semantic properties (Pitt 2013). But even this broad interpretation is circumvented on at least two counts. The first is practical and concerns the requisite conditions for successful performance on the edge. Skilled sportspersons in high performance scenarios are particularly responsive and attuned to the situation and what the environment calls forth. Running an explicit mental conversation on what one is about to do is a sure way to botch the attempt. Even if sportspersons and performers often give gripping narratives of their exploits, or in some sports they may pause to deliberate their next move, propositional content and linguistic articulation should be absent when they need to be *fully* engaged. In the thick of the action, as a gymnast lunges to grip the parallel bars or a high cliff diver jumps off the platform, the moment for any kind of representational content that converts semantically is either premature or well past. Put otherwise, when semantic content takes over, it obstructs the path to flow or *mushin* states.

There are two sides to this obstruction. One concerns processing speed, the other attentional focus. Concerning the first, it seems likely that at the neural level an enactive stance affords more efficiency because it is less constrained in terms of neural resources. From a purely practical standpoint, less processing is more efficient. As Rodney Brooks commented regarding classical computational models and their quest for AI, 'When we examine very simple level intelligence we find that explicit representations and models of the world simply get in the way.' (1991, 140)[10] Of any two sportspersons, if one can eliminate the processing required for internal representations and still respond appropriately to the situations she encounters, she will be faster. In martial arts, particularly swordsmanship where finely split seconds determine outcomes, this advantage in timing marks the difference between surviving and being cut down.

As we argued above, many normal human activities seem to require mental representation. If sportspersons are able to train themselves to avoid or move beyond processing that requires representations, they reduce the need for cognitively intensive resources. This reduces limitations due to processing time. In some aspects then, they function in ways that are similar to basic organisms, but at the same time they maintain higher level capacities, some distinctly human.

This is where attentional focus comes to the fore and enactivism proves superior not only to traditional cognitivism but also to alternative non-representational models. Dreyfus (1972) and those who follow in his stead rely on absorbed coping. But, just as computational views, this problematically shifts skilled action to the subpersonal level. Thus, the 'magic' of athletes or martial artists, and how they quiet their 'mental chatter,' is explained by hiding the process inside a black box. Yet, skillful sportspersons and performers are clearly in control and highly aware—particularly in flow and *mushin* states. Contrary to this, an enactive model *transparently* accounts for the focused attention of skilled performers, as the ensuing will argue. Evidently, without semantic or representational content, there is no mental chatter to speak of (or other processes that may lead to distraction). This leads to the second way to circumvent the semantic view of mental representation such that high performing states are better explained.

This way is philosophical rather than practical: it is concerned with the concept of representation directly, and suitably expands on the theoretical underpinnings of the previous point. When a big wave surfer or a fighter face do-or-die situations, their intentionality and engagement with such situations do not involve properties like truth, veridicality, consistency, accuracy, or appropriateness, all of which explain intentionality for the representational theory of mind. In other words, as they act, immersed in the moment, the movements that work are not evaluable in terms of truth, veridicality, or any of the other aforementioned conditions obtaining. Recall that for Hutto and Myin, 'Our basic ways of responding to worldly offerings are not semantically contentful' (2013, 82). It is not that *this* movement is true or accurate or consistent in a semantic sense, as opposed to another one. Rather, there is an array of possible movements, which the sportspersons' kinetic repertoire makes available, that variously fit the specific scenario they face (Ilundáin et al. 2017). There is neither cognitive room nor need to entertain such semantic properties in actual performance, either overtly or implicitly. Overtly as adduced, representation of explicit articulation compromises success. Implicitly and subpersonally (if considering a subconscious account of mental representation more agreeable to cognitive neuroscience and sports psychology), besides being ineffective, such semantic concerns are superfluous to handle the kinetic problems of high-level performance. In short, the fundamental manner in which the sophisticated actions and states proper of flow and *mushin* operate is enactive rather than either cognitively contentful or reliant on Dreyfusian mindlessness.

Cognitive emptiness then bypasses semantic content and also vacates distracting thoughts, which facilitates the sort of intense focus needed. Concisely put, enaction is the way to best 'invite' success. This contentless cognition also includes, as seamless and integrative performance, emotional equanimity and volitional effectiveness: sportspersons operate without distracting thoughts, worries, or misgivings. But, how do these contentless states help performers *specifically*?

When Alex Honnold is free soloing and onsighting a big wall—climbing it without any ropes or gear other than chalk and shoes nor prior reconnoitering—he is performing one of the most complex tasks a human can do in an actual life or death situation. Since the stakes could not be higher, literally and figuratively, this engages him wholly from an enactive perspective: not only does he figure out which line to take up the wall, but he calms himself to remain in control in a situation where, past a certain point in the climb, a wrong move would mean certain death. In other words, Honnold integrates rational assessment, emotional control, skillful climbing prowess, volition, and even imagination[11] into a seamless performance. Because such climbs last for hours, he is not 'flowing' for the duration. Rather, he alternates among different cognitive engagements, now relying on deliberation, now on visualization, sometimes utilizing overt semantic content. But, when he reaches what he calls the crux of a climb, he focuses intensely; then it is just him and the rock (Honnold and Roberts 2016). When, after pondering a difficult move, he gets going, his movements respond immediately and resonantly to the rock's surface and his kinesthetic and proprioceptive dynamics. At that point, there is no room to rely on representational content of the kind that would have Honnold ask himself 'should I reach for that hold or not?' There is only climbing. The seamless yet controlled action—subject to kinetic judgments—that an enactive model of cognition affords, without intermediaries that impede processes, is most necessary.

This cognitive, non-representational emptiness also animates the master swordsman when he controls an opponent in a life or death duel. Zen master Takuan Sōhō evokes this in his letter to famed samurai and *kenjutsu* master Yagyū Munenori,

> Completely oblivious to the hand that wields the sword, one strikes and cuts his opponent down. He does not put his mind in his adversary. The opponent is Emptiness. I am Emptiness. The hand that holds the sword, the sword itself, is Emptiness. (1986, 37)[12]

Like Honnold, this is most advantageous in a life or death situation; it calls for complete attention and no distraction. This is achieved through an emptiness that vacates content—semantic content and other mental representations whether explicit or implicit. Such contentful 'ways' have long been grounded in fully enactive processes; it should be emphasized, as noted, that these are holistic and integrated processes—there is no gap between the cognitive and the bodily in high performance. There are two key aspects to this, facilitated by the just discussed semantic 'emptiness': immediacy of response, and context-sensitivity and adaptability.

Immediate response in the sort of high performance scenarios in which *mushin* and flow operate is peremptory. Unsurprisingly, such immediacy is predicated on the absence of contentful states. The performers do not have time to articulate in *real time* a plan of action. Moreover, as stated, there are no truth or veridicality or satisfaction conditions that obtain. To illustrate such immediacy, consider sumo (*sumō*, 相撲) wrestlers facing off. Crouching very low, they gaze intently on each other. At this moment, the ability to react simultaneously and instantaneously to the opponent's move is crucial. But, this is so only when rivals initiate a *veritable* move, for oftentimes they show the faintest trace of a move in order to trigger a faulty response that disconcerts adversaries. Merely pondering whether to do a *hazuoshi* (筈押し), hooking opponents under the armpits, or a *henka* (変化), sidestepping the antagonists' charge, *will* cost them the bout, as they will react too slowly. Rather, there has to be an in-the-moment matching to rivals, much like a surfer adapts to the changing wave's surface. Wrestlers (and sportspersons) need to have an immediate and specific attunement to the situation as they adapt to *particular* opponents (or environmental feature such as a wave or rock face). That is, they fight s*pecific* rivals in a given moment of form, not just 'other fighters.' This highly specific sensitivity to context and adaptation is predicated on an enactive model. It is precisely because of the kind of attuned resonance between situations and skills that characterize such high performances that absence of representation gives the advantage.

4. What Are the Advantages of the Enactively Embodied Approach?

It is not only possible but even advisable to explain high performance without appealing to mental content. The ensuing lays out how enactivism offers theoretical and practical advantages as an explanation of performance in sports, martial arts, and other embodied practices. It also explains why sports and high performance validate enactivism.

To begin with the theoretical aspect, enactivism is more parsimonious than mainstream alternatives. It eliminates superfluous explanatory elements, whether these be mechanisms (physiologically grounded or conceptually deduced) or ancillary representational factors. Conceptually, this model better captures the underlying conditions of our intentional states as expressed in the type of performance typical of flow and *mushin*. Rather than semantic properties, we find immersed, adaptable, and highly context sensitive non-representational engagements. Additionally, this is expedient because such simplicity can account for the immediacy that high-stakes action demands. Relatedly, this also accounts for the reaction speeds proper of fast-paced sports, which need to be performed under control and not merely automatically (Ilundáin et al. 2017).

This is exciting theoretically because, besides rethinking from the ground up how to conceive highly skilled performance, the enactive account as developed here is also compatible and makes advantageous connections with other theoretical approaches that bring mutually complementary perspectives. Two such programs are dynamical systems theory and ecological psychology. In dynamical systems theory, the dynamic adaptations of developing (Thelen and Smith 1995) and mature organisms and persons (Kelso 1995) result in coordinated action that can be congruently explained via the scaffolding of representational content atop contentless cognition and skills. Likewise, ecological accounts of performance (Christensen et al. 2016; Davids et al. 2015), when these do not rely exclusively on representations at a fundamental level, are also compatible and offer mutual benefits given that the breadth of the ecological model accounts for the complex multi-level engagements of experts across various activities while the enactive approach provides fine-tuned transparency to said engagements.

To discuss the practical aspect now, sports and similar types of high performance validate and enrich the enactive account as they extend its domain of applicability from basic minds to skilled contexts where highly sophisticated cognition is involved. Such activities at that level of expertise are some of the most taxing and refined. Moreover, they engage agents *holistically*: the performers need to be attuned to the environment holistically, that is, integrating to the utmost degree intellectual, emotional, volitional, kinetically skillful, and other capabilities. In this way, this is applicable to the whole gamut of explanatory levels—from microscales to meso and macroscales. In a related manner, this approach aligns with empirical research that corrects all too often invoked but erroneous maxims such as the one about keeping the eye on the ball (Knudson and Kukla 1997). Finally, this also has pedagogical advantages. An enactive account supplements traditional explicit articulation and verbalization of techniques. Thereby, enactivism affords creative enactive and embodied non-verbal interactions among coaches and sportspersons while the activity is carried out in the pertinent environment in playful ways that push performers to hone their skills (Ilundáin-Agurruza 2017a). For example, rather than predictable drills and explicit instruction, trainees can be presented with kinetic problems that need collaborative engagements without verbalization.

5. Conclusion

When superb tennis players or aikidoka face off against opponents, mental representation has been the standard explanation for their skilled actions in sport psychology, classical computational accounts of the mind, and sport philosophy. Alternative accounts that disposed of representations, such as Dreyfus' and akin phenomenological models, have proven inadequate. In this article, we have argued for an enactivist stance, specifically Hutto and Myin's radical enactivism (REC). If basic minds embody the paradigmatic case for REC, we have argued that the highly skilled performances of sportspersons and others, particularly those characterized by flow and *mushin* states provide a better account on several levels. It affords a more parsimonious and transparent explanation of highly performance in the thick of action. Its semantically contentless cognition not only prevents mental chatter and enables greater attentional focus, but also helps elucidate the immediacy of response, context-sensitivity, and adaptability characteristic of such performances and states. The outcome is a number of theoretical and practical advantages that include complementary theoretical affinities with other disciplines and an expedient holism that may lead to supplementary and beneficial techniques. All in all, the enactive approach and high performance in sports are well matched. They deliver on their promises and, we surmise, augur even richer insights and better performances if theory and practice are further developed along the lines presently discussed.

Notes

1. 'Bodyminded' here refers to a holistic integration of psychic and somatic elements such that dualist dichotomies are sidestepped (Campos 2014; Ilundáin-Agurruza 2016, 2017a). It follows a philosophical tradition that, in the West, is best represented in the writings of pragmatists (Dewey, Peirce) and phenomenologists (Merleau-Ponty, Ortega y Gasset).
2. But see Hutto (2015) for an enactive account of imagination without representations, and Ilundáin-Agurruza (2017b) for a phenomenological and enactive model that also disposes of representations in cases of imaginings that involve performance.

3. The following discussion on representation reflects the preponderant dualist framework and vocabulary of mainstream and traditional views on the matter. The model presently developed rejects such framework.
4. It is worth noting that neuroscientists specifically address the question of how neuronal firing patterns give rise to our actions. Neuroscientific accounts of representation range from simpler views that still tie these structures to images of the world (Damasio 1999) to more sophisticated frameworks where action plans select motor representations from the stock of available schemas needed to respond to internal and external cues (Jeannerod 1997). The important point is that questions concerning whether such structures really represent or refer to features of the world are far less important to neuroscientists than philosophers.
5. Davidson (1982) and Stich (1979) for example have both argued that conceptual content depends on the ability to use and speak a complex language.
6. In the case of nature sports, one reason for this might be that because athletes interact with such powerful features (waves, mountains, etc.), they need to be able to work with, rather than against those features. This often means 'giving oneself over,' or surrendering, in a sense. For a detailed discussion of the interaction between sportspersons and natural features in nature sports, see Krein (2014).
7. His description should not *necessarily* be understood as a mindless and absorbed coping where the sportsperson is unawsare (see Breivik 2013 for a critique of absorbed coping).
8. Krein (2015) discusses this point as it relates to questions of how nature sports change when they are adapted to fit into formal competitive frameworks.
9. Considering the standard Buddhist claim that views our experience as deluded, in particular because we see the through the delusion that the self or ego exists, it might follow that the experience of no-mindedness fits well with the idea that content and consciousness are themselves delusions.
10. This is one reason that many AI researchers began to abandon their commitment to building representations into their systems.
11. Such imaginings are not necessarily representational. In fact, in such high performance scenarios the imagination is also enactive (see endnote 2 for references).
12. The translation is slightly misleading, as the original Japanese word rendered as 'mind,' *shin* 心, is better translated as 'heart-mind' given its deep embodied, psychosomatic connotations.

Disclosure Statement

No potential conflict of interest was reported by the authors.

Funding

This work was supported by the Australian Research Council Discovery Project 'Minds in Skilled Performance' [grant number DP170102987].

References

ACKER, R.B. 1998. *Kyudo: The Japanese art of archery*. Boston, MA: Charles E Tuttle Publishing.
BREIVIK, G. 2013. Zombie-like or superconscious? A phenomenological and conceptual analysis of consciousness in elite sport. *Journal of the Philosophy of Sport* 40 (1): 85–105.
BROOKS, R. 1991. Intelligence without representation. *Artificial Intelligence* 47 (1–3): 139–159.
CAMPOS, D. 2014. On creativity in sporting activity: With some consequences for education. *Fair Play. Revista de Filosofía, Ética y Derecho del Deporte*. 2 (2): 52–80.
CHRISTENSEN, W., J. SUTTON, and D. MCILWAIN. 2016. Cognition in skilled action: Meshed control and the varieties of skill experience. *Mind & Language* 31 (1): 37–66.
CSIKSZENTMIHALYI, M. 1990. *Flow: The psychology of optimal experience*. New York, NY: Harper and Row.

DAMASIO, A. 1999. *The feeling of what happens*. San Diego, CA: Harcourt.

DAVIDS, K., D. ARAÚJO, L. SEIFERT, and D. ORTH 2015. Expert performance in sport: An ecological dynamics perspective. In *Routledge handbook of sport expertise*, edited by J. Baker and D. Farrow. London: Routledge: 130–144.

DAVIDSON, D. 1982. Rathional animals. *Dialectica* 36 (4): 317–327.

DENNETT, D. 1987. *The intentional stance*. Cambridge, MA: MIT Press.

DREYFUS, H. 1972. *What computers can't do*. Cambridge, MA: MIT Press.

ERIKSEN, J. 2010. Mindless coping in competitive sport: Some implications and consequences. *Sport, Ethics and Philosophy* 4: 66–86.

HONNOLD, A. and D. ROBERTS 2016. *Alone on the Wall*. New York, NY: W.W. Norton & Co.

HUTTO, D. 2015. Overly enactive imagination? Radically re-imagining imagining. *The Southern Journal of Philosophy* 53: 68–89.

HUTTO, D., and E. MYIN. 2013. *Radical enactivism: Basic minds without content*. Cambridge, MA: MIT Press.

ILUNDÁIN AGURRUZA, J. 2016. *Holism and the cultivation of excellence in sports and performance: Skillful striving*. London: Routledge.

ILUNDÁIN AGURRUZA, J. 2017a. A different way to play: Holistic sporting experiences. In *Philosophy: Sport. Macmillan Interdisciplinary Handbooks*, edited by R. Scott Kretchmar. Farmington Hills, MI: Macmillan: 319–343.

ILUNDÁIN AGURRUZA, J. 2017b. Muscular imaginings: A phenomenological and enactive model for sports and performance. *Sports, Ethics and Philosophy* 11 (1): 92–108.

ILUNDÁIN AGURRUZA J., K. KREIN, and K. ERICKSON. 2017. Excellence without mental representation: High performance in risk sports and Japanese Swordsmanshi In *MIT handbook of embodied cognition*, edited by M. Cappuccio. Cambrige, MA: MIT Press.

JEANNEROD, M. 1997. *The cognitive neuroscience of action*. Oxford: Blackwell.

KELSO, J.A.S. 1995. *Dynamic patterns: The self-organization of brain and behavior*. Cambridge, MA: MIT Press.

KNUDSON D. and D. KUKLA. 1997. The impact of vision and vision training on sport performance. *Journal of Physical Education, Recreation and Dance*: 17–24.

KOTLER, S. 2014. *The rise of superman*. New York, NY: Houghton Mifflin Harcourt.

KREIN, K. 2014. Nature Sports. *Journal of the Philosophy of Sport*. 41 (2): 193–208.

KREIN, K. 2015. Reflections on Competition and Nature Sports. *Sport, Ethics and Philosophy*. 9 (3): 271–286.

KREIN, K., and J. ILUNDÁIN-AGURRUZA. 2014. An East-west comparative analysis of *Mushin* and flow. In *Philosophy and the Martial Arts*, edited by G. Priest and D. Young. London: Routledge: 139–164.

KRIEGEL, U. 2014. *Current controversies in philosophy of mind*. New York, NY: Routledge.

MILLIKAN, R. 1984. *Language, thought and other biological categories*. Cambridge, MA: MIT Press.

MOE, V.F. 2007. Understanding the background conditions of skilled movement in sport: A study of searle's 'background capacities'. *Sport, Ethics and Philosophy* 1: 299–324.

NOË, A. 2004. *Action in perception*. Cambridge, MA: MIT Press.

O'REGAN, J., and A. NOË. 2001. What it is like to see a sensorimotor theory of perceptual experience. *Synthese* 129: 79–103.

PAUPORTE, A. and F. PAUPORTE (PRODUCERS) and F. PAUPORTE (DIRECTOR). 2007. *Lines* [DVD] Quinta Films.

PITT, D. 2013. Mental representation. In *The Stanford encyclopedia of philosophy*, edited by E.N. Zalta. https://plato.stanford.edu/archives/spr2017/entries/mental-representation

ROWLANDS, M. 1997. Teleological semantics. *Mind* 106 (422): 279–303.

SHAPIRO, L. 2011. *Embodied cognition*. London: Routledge.

SHEETS-JOHNSTONE, M. 2009. *The corporeal turn: An interdisciplinary reader*. Exeter: Imprint Academic.

SHEETS-JOHNSTONE, M. 2011. *The primacy of movement*. 2nd ed. Amsterdam: John Benjamins.

STICH, S. 1979. Do animals have beliefs? *Australasian Journal of Philosophy*. 57 (1): 15–28.

THELEN, E. and L. SMITH 1995. *A dynamic systems approach to the development of cognition and action*. Cambridge, MA: MIT Press.

THOMPSON, E. 2007. *Mind in life: Biology, phenomenology, and the sciences of mind*. Cambridge, MA: Harvard University Press.

Neuropsychology Behind the Plate

Jordan Edmund DeLong

ABSTRACT
In baseball, plate umpires are asked to make difficult perceptual judgments on a consistent basis. This chapter addresses some neuropsychological issues faced by umpires as they call balls and strikes, and whether it is ethical to ask fallible humans to referee sporting events when faced with technology that exposes "blown" calls.

Introduction

'I never questioned the integrity of an umpire. Their eyesight, yes.' *Leo Durocher* (Light 2005, 303)

Baseball in America wouldn't be the same without heckling. While an underperforming player may become a target for insults, an umpire receives universal scorn and disrespect. Michael Tolley, the creator of the online insult repository www.heckledepot.com, stated that heckling in baseball is 'as much a part of the game as $5 hot dogs …' (McGovern 1999). Many of the top-rated umpire-directed heckles on Tolley's website center on visual impairment, including 'Lenscrafter called … they'll be ready in 30 min.', 'We know you're blind, we've seen your wife', and 'When your dog barks twice, it's a strike!' (Tolley n.d.) This impulse to tear down (or tear apart) umpires isn't unique to baseball, and can lead us to question what practical and ethical obligations players, umpires, and even sports fans have when trying to enforce the rules of our games. Previous work has investigated what it really means when an umpire makes a call, and what conflicts arise when that call is bad (Bordner 2015; Russell 1997). This paper will attempt to show how, from a neuropsychological perspective, it is truly impossible for umpires to live up to the standards we set for them, and that perhaps the lack of precision is exactly what makes games entertaining.

Performing under fire is nothing new for umpires, who are elite-level perceptual performers on par with the athletes on the field. The process of becoming an MLB umpire is arduous, and starts by attending a paid umpire school. After graduating near the top of their class, new umpires begin their career working in the lowest levels of professional baseball, progressing through each level of the minor leagues including Rookie, Class A Short Season, Class A, Double-A, and finally Triple-A (Rogers 1999). Even at the lower levels of the sport, umpires must show an ability to manage games, call pitches, and enforce all of baseball's myriad rules at each level. Those that cannot improve their accuracy within three years are dismissed. If an umpire manages to make it through to the Triple-A ranks, they have earned the right to be considered for a position in the major leagues. Many umpires do not make it past this stage, however, because turnover in the major league is so only about one opening per season. While the 68 Major League Baseball (MLB) umpires are well compensated, minor league umpires work for anywhere between $1900 and $3500 a month (MiLB.com n.d.). In a real sense, the path to becoming an MLB umpire requires 8–10 years of consistently high perceptual performance and low wages that create a pathway that is more selective than Juilliard, Harvard Medical School, or the US Navy SEAL training program.

Despite rigorous training and all their years of experience, major league plate umpires are incorrect in roughly 10–15% of their pitch calls behind the plate (Kim and King 2014; Moskowitz and Wertheim 2011). While incorrect calls have always bothered fans of the sport, the ability to easily scrutinize pitch calls grew considerably in 2008 when the MLB installed the PITCHf/x tracking system into every ballpark in baseball, a system which uses a set of cameras to track the location of a pitch 60 times per second (DiMeo 2007). PITCHf/x collects the velocity and trajectory of every pitch from every game, and has been integrated into television broadcasts directly, allowing viewers at home to see the exact location of the pitch relative to the strike zone in real time. In the past, fans had to use their own perception of the strike zone to lambast umpires, but now a blown call is broadcast for all to see in vivid HD. In a cruel twist of fate, modern video replay rules consider the calling of a ball or strike a judgment call by the plate umpire, and are not eligible for video review per MLB Rule 9.03a. This is a blow to the authority of the umpire, exposing the uncomfortable truth that a superior viewpoint and years of expert training still can result in a blown call (Collins 2010). Put plainly, umpires could get away with a lot more before the adoption of PITCHf/x, but their job isn't only getting harder because of increased transparency but also due to the performance of modern players.

Today's players are tracked using an impressive battery of performance metrics, but baseball's history extends so far into the past that modern statistics we take for granted, like the miles-per-hour of a given pitch, are simply not available. Given that baseball developed before the majority of US households had running water or flush toilets, it isn't surprising that there are few accurate estimates of pitch speed during baseball's first half century. This makes comparison of players from these early eras difficult, but some isolated examples exist. One of the earliest tests was conducted in 1917 when Hall-of-Famer Walter 'The Big Train' Johnson was at the height of his powers—the year before he would win his second of three Pitching Triple Crowns by leading the league in wins, strikeouts, and having the lowest earned run average (ERA). Scientists at a munitions laboratory in Connecticut recorded his fastball as reaching 91.36 miles per hour (Associated Press 1939). This speed, while considered dominant for the time, is below the average pitch speed of 92.0 miles per hour for the 2015–2016 season, including only pitchers that threw 30 or more innings (FanGraphs.com 2016). This is not to claim that Johnson couldn't cut it in the big leagues today, but simply to point out that the average speed of pitches was slower at that time. There are many possible reasons for this difference, including the fact that relief pitching, the act of swapping in a new pitcher once a starter shows signs of fatigue, had yet to be adopted to stave off injury and keep an arm 'fresh'.

In previous eras of baseball, pitchers would often play games on back-to-back days and throw at a deliberately slower pace to preserve stamina. Today's pitchers rarely play entire games and typically take at least five days to recover after each game started. Modern pitchers can push themselves to 'leave it all on the field', rather than worry about performance the next day or even the next inning. The difference in velocity produced by MLB pitchers then and now means that the task of both modern batters and umpires is more difficult now than it was in the early days of baseball. One indication of this is that all of the all-time career batting average leaders had all retired from baseball before World War II, with the notable exception of Ted Williams, who interrupted his baseball career to serve in both World War II and Korea, and didn't quit baseball until 1960 (Baseball Reference n.d.). Perhaps the real fireballers of yesteryear could go toe to toe with today's stars, but the depth of pitching talent in the modern bullpen would have been unthinkable in the early twentieth century.

With increased knowledge of the biological and physical aspects of throwing a baseball, pitchers regularly throw harder than was once thought to be physiologically possible. Even with a decreased pitch count and increased time for recovery, in today's MLB it is rare for a pitcher to play for any length of time before blowing out the ligament in their elbow and undergoing Tommy John surgery. Joe Kelley, starting pitcher for the Saint Louis Cardinals, explained that 'In the end, Tommy John is like death. It's going to get you' (Goold 2014). In a real sense, pitching in major league baseball regularly surpasses the physical limitations of the human body, straining ligaments to the brink.

Amazingly, the neurological and decision-making limitations facing umpires and batters at the other end of sixty feet and six inches are far more complex than the physical limitation of pitching arms, and involve a series of perceptual and cognitive processes that are not yet fully understood. Although our knowledge of how to maintain, augment, and repair a pitcher's arm has increased over time, we have not made comparable strides in how to improve perception and decision-making. In a real sense, the brain of an umpire has changed little since 1871 and is forced to contend with twenty-first century technology. Our brains, despite their impressive computational abilities, are fundamentally limited in ways most people do not recognize.

The Eye, Saccades, and Neural Transmission Rate

Most people experience the world around them as a stable, static perceptual experience. While our visual world feels like it is perfectly rich and detailed, studying the basic structure of the eye reveals that our view of the world is a construction which is cobbled together by our brains. The retina only supports full-resolution viewing in the very center of our visual field in an area called the fovea, and visual acuity drops off rapidly toward the periphery. Our eyes move to keep this sensitive area focused on different important areas of the world in front of us, gathering new information two to three times per second every moment that we are awake. When an object moves, our eyes either snap from one location to another or follow an object using a reflex that allows for smooth pursuit, effectively allowing us to 'keep our eye on the ball'. Both types of eye movements have their drawbacks, as a movement that abruptly jumps from one location to another induces momentary blindness during the transition and smooth eye movements work only when the object viewed moves slowly enough to be tracked on-the-fly.

Instead of moving their eyes to track an action, research in the laboratory and the field suggests that the best performers 'lock-on' to a specific target or and keep a 'quiet eye' before performing an action, meaning that they fixate their eye earlier and linger longer than average (Vickers and Adolphe 1997). In a study which tracked the eyes of umpires, researchers found that experts set their gaze earlier and lingered longer than near-experts, suggesting that umpires who develop the ability to keep their eyes locked in a position, typically the release point of a pitcher, may be more successful than those attempting to follow the ball (Millslagle, Hines, and Smith 2013). This skill was also found in batters, who have a tendency to keep their eyes quiet but to make tracking motions with their heads when trying to hit the ball (Shank and Haywood 1987). Simply tracking the ball isn't enough, however, as batters are regularly faced with predicting where and when the ball will travel, a task made even more difficult with pitches that may rise, fall, break, or arrive off speed.

While umpires get a reprieve from having to understand when the ball arrives at the plate, they are held to a high standard when proclaiming where it arrives. The visual system of an umpire faces the same basic issue that all humans face each day—making sense of a three-dimensional world given two-dimensional input. While the PITCHf/x system can utilize several high-resolution cameras strategically placed at divergent angles around the ballpark, umpires are forced to analyze information from their position behind the catcher. In addition to the demands of tracking the three-dimensional flight of a ball utilizing a single viewpoint, umpires are forced to contend with the raw speed generated by pitchers. In calculations of speed, the average pitch in major league baseball will travel from the pitcher's hand to home plate in approximately 400 ms (Adair 2002). If it wasn't difficult enough to form a perceptual estimate after viewing such limited input, batters, catchers, and umpires are forced to contend with the questionable architecture of the visual system.

In the primate brain (along with many other mammals), the most powerful region associated with visual processing is located far away from the eyes at the back of the head. This quirk of cerebral real estate means that it takes any visual signal additional time to travel to the back of the head for processing. In macaque monkeys, this signal takes on average 77 ms to reach the back of the monkey head (Nowak, Munk, Girard, and Bullier 1995), and will take longer inside our bigger human skulls. In a real sense, this means that we live roughly 100 ms in the past, though most people never feel like their vision lags behind the world. Recent research argues that this is because our brain may be projecting our perception forward in time, simulating what we expect to happen and revising our perception after an event occurs (Maus, Ward, Nijhawan, and Whitney 2012).

Filling in these temporal gaps is nothing new for the visual system. Every time our eyes jump from location to location, the visual signal is suppressed (Volkmann 1986). This leaves a gap in time, which our brain fills in retroactively in a phenomenon highlighted in the 'Stopped Clock Illusion' (Yarrow, Haggard, Heal, Brown, and Rothwell 2001). In this illusion, our brains fill in the time between saccades with an image, backfilling the perceptual gap with an image of where the saccade landed. This effect may be responsible for pitch framing, a technique where catchers quickly move a ball that is pitched outside the strike zone back inside without the umpire noticing.

Recent research into pitch framing has shown that catchers have a wide range of talent when deceiving umpires, but that a successful pitch framer can often earn their pitcher a slight but significant edge (Pavlidis and Brooks 2014). By plotting where an umpire's typical pitch boundaries are located, it is possible to see how far a catcher can extend the strike zone. What makes one pitch framer better than another is still under debate in the baseball literature, however what we know about how our brain backfills images can help guide our theory. With an umpire having to move their eyes, we know that there exists a short, roughly 100-ms window of opportunity where the catcher may be able to change the position of the ball with a brief, slight movement and trigger the umpire's brain to reinterpret the pitch with this new information. This ability, not unlike a magician or pickpocket's ability to distract, could be trained in catchers by making them privy to the short time when umpires may be distracted by eye movements.

Although pitch framing has become an immediately accepted part of baseball, it poses ethical issues to the game. While promoting pitch framing, general managers are trying to gain an advantage through a catcher's ability to exploit psychophysiological deficits in the umpire's perceptual system. By providing this advantage, catchers are essentially breaking the rules of baseball. From a fairness perspective, should teams receive a small but significant edge because some players are better at breaking the rules? This also poses stronger issues when arguing from formal theory, especially when considering Bernard Suits' argument that not playing a game according to the rules functionally means you aren't playing the game at all (2014). Clearly, we can sidestep this issue by saying that the umpire's ontological authority makes all debate fruitless, but in a PITCHf/x era the disparity between what is called and what might have been called has never been more transparent (Collins 2010).

When catchers pitch frame, they are deliberately trying to break the rules to their favor, but is this an injustice? In a sense, catchers from opposing teams also can attempt to tip the scales in their favor, so there is little outcry from fans. Similar conscious rule-breaking happens in other sports as well, and typically isn't considered to invalidate the game. In basketball, fouling a shooting player is clearly against the rules; however, fouls are used strategically in a variety of situations. A 'good' foul may occur when it is preferable to send a player to the foul line rather than to allow them to hit a routine shot. At the end of a game, it may be advantageous for the trailing team to commit fouls rather than allow time to expire from the play clock. Breaking a rule doesn't immediately mean that a game is invalid or an injustice has been committed, but smaller, clearer injustices influence sporting events on a daily basis.

Judgment and Bias

Although umpires face basic perceptual issues, there are yet-more difficulties in calling balls and strikes. Even if an umpire had bionic, PITCHf/x-like perceptual precision, would they be 100% accurate with their pitch calling? Modern research in judgment and decision-making suggests that even if umpires were gifted with perfect sight, they may still be influenced by their own beliefs and biases. This phenomenon isn't specific to umpires; human beings have been shown to possess implicit biases that can shape our behavior even when we are unaware of their influence (Greenwald and Banaji 1995). In recent years, research into implicit racial bias has shown that people may make biased decisions based upon stereotypes of race (Fazio, Jackson, Dunton, and Williams 1995; Greenwald, McGhee, and Schwartz 1998), gender (Nosek, Banaji, and Greenwald 2002), and disability (Strohmer, Grand, and Purcell 1984; Thomas, Doyle, and Daly 2007), without any awareness of their own bias.

While an umpire is expected to provide a stable strike zone, they cannot help but be influenced by the game situation at hand. For example, if a batter has two strikes, an umpire will be far less likely to call a borderline pitch a strike (Green and Daniels 2014). Similarly, when a pitcher is about to walk a batter, an umpire's strike zone will expand, meaning that they are more likely to call the next pitch a strike. Another bias extends into the postseason, when the strike zone has been shown to grow, favoring pitchers. Luckily, while these types of bias may be pervasive across baseball, teams should be impacted roughly equally. Unfair treatment may come into play when umpires are faced with other types of bias that target teams differently.

During the 2015 MLB season, Chicago Cubs manager Joe Maddon was ejected from a game for arguing the strike zone with plate umpire DJ Rayburne. While having a manager (especially Maddon) ejected from a game is not an uncommon occurrence, Maddon's criticism of the strike zone was directed specifically at Rayburne's bias to favor a number of veteran players from St. Louis over the rookies that batted in the Cubs lineup. While Maddon's on-field rant cannot be written in polite company, he summed up his argument post-game by saying 'You play a veteran club with a veteran battery and you have guys that barely have a month in the big leagues. I'm not going to take it. Our guys deserve equal treatment' (Gonzales 2015). Maddon is describing the Matthew Effect (Merton 1968), a rich-get-richer bias that appears in nearly every area of human decision-making. This effect was shown to influence the intellectual development of students, where those labeled as 'slow readers' receive less instruction, read less, and eventually become self-fulfilling prophecies.

Research conducted by Jerry Kim and Braden King have analyzed PITCHf/x data that supports Maddon's claim that umpires implicitly reward and punish players. By looking at 756,848 pitch calls from the 2008 and 2009 MLB seasons, it was observed that plate umpires tend to change their strike zone for a variety of different reasons, with large changes observed in things like player status and smaller changes related to conditions like the race of the player (Kim and King 2014). In a real sense, all-star pitchers are more likely to get the benefit of the doubt over unestablished players. It will be interesting to see how the Chicago Cubs players are treated now that they have at long last won the World Series and are no longer considered 'cursed'.

This form of bias is more than just an officiating phenomenon but is behind many of the cues of quality we take for granted today. While a bias against certain players certainly seems unfair, human decision-making follows several tried-and-true heuristics which may introduce bias in some situations, but in our day-to-day life help accurately guide our decisions with a minimum of effort. The fact of the matter is that all-star players typically do perform better than their less-celebrated counterparts, so it isn't unreasonable for an umpire's brain to assume that they are indeed performing better, even if they aren't. This same process happens when we decide to purchase brand name vs off-brand products, often opting to pay more for a label that we know and trust (Park and Lessig 1981).

While our intuition may suggest that errors in perception and errors in judgment are fundamentally different, the gap between our eyes and our hearts may not be nearly as distinct as we would like to believe. It may be the case that our perception is far from insulated from our social and physical environment, and the world around us may literally change the way we perceive the world around us (Bruner 1957, 1992). Students asked to walk up a hill estimate the slope of the hill as being greater when asked to carry a heavy backpack up it than when not carrying a backpack (Bhalla and Proffitt 1999). Golfers who are performing well report seeing the size of the hole as larger compared to those that are struggling (Witt, Linkenauger, Bakdash, and Proffitt 2008). Even baseball players appear to see the ball differently based upon their play (Witt and Proffitt 2005), and even have commented on this phenomenon; Mickey Mantle reporting after a home run that the ball was 'as big as a grapefruit'.

Evidence is mounting that our visual perception is influenced by our physical and psychological circumstances. It also appears that we see what we want or expect to see, and umpires are no exception. Research has shown that when given an ambiguous image, people see it in the way that serves them at the time of perception (Balcetis and Dunning 2006), which could explain why two opposing fans may see the same replay and both conclude it is conclusive in opposite ways. Umpires are influenced to see the ball as being either in or out of the strike zone based upon their situation including the pitch count, the players involved, and the stakes of the game.

Despite the demands of fans, pitch-callers are still human and face the same biases and limitations we share. Behavioral scientists know that their bias can contaminate an experiment without awareness, and great pains are taken to maintain validity by keeping subjects and experimenters as blind as possible to condition. There is no good way to keep an umpire blind, no pun intended. A plate umpire, who is disinterested in the game of baseball, compartmentalizes calling balls and strikes, and isn't influenced by game situations wouldn't be human. In a real sense, it will be much easier to build a machine that calls strikes than fundamentally change the foibles of human cognition and perception.

Building a Better Pitch Caller

If a plate umpire is handicapped by their fundamental neuropsychological limitations, what can we do to improve their performance? One solution may be to have them begin their training at a younger age to take advantage of the inherent neuroplasticity of youth. While MLB umpires have worked through many years of intense practice, it might be possible that shifting their training into childhood could pay dividends. While many children begin playing baseball from a young age, umpires typically begin their training after high school or college at a time where their brain already has started to exhibit less neuroplasticity. While adult brains can create new neural pathways, the flexibility found in a child's brain allows for much more adaptation. Therefore, children can survive and recover from having half their brain surgically removed, an operation that no adult could recover from. It's been hypothesized that the reason Michael Jordan couldn't cut it in minor league baseball was because his brain wasn't young enough to re-wire itself for elite-level baseball play despite having the body of an elite-level athlete (Klawans 1996). It seems unlikely that fathers will prefer buying their children their first umpire mask over their first glove, but early training could produce better umpires.

Other possibilities for improving umpire performance are to change physical aspects of the field, a strategy that the MLB has adopted in the past. Around the turn of the twentieth century, Amos 'The Hoosier Thunderbolt' Rusie vexed hitters with his blistering speed and slightly spotty control. Rusie prompted the MLB to move the mound backward from fifty feet from home to sixty feet and six inches in 1893 when one of Rusie's wild pitches put a future hall of famer in a coma for four days. While moving the mound back did decrease strikeouts and increased hits across the league it didn't deter Rusie—the year after the mound was moved back he still led the league in wins, starts, shutouts, all while maintaining a startlingly low ERA of 2.78, roughly half of the league average of 5.33. Eventually Rusie's career would be cut short by contract disputes, arm troubles, and eventually poor play, but his impact on the game of baseball remains to this day.

In later years, two major changes lead to a period of pitching dominance. The first was the strategic adoption of relief pitching and the second was an expanded strike zone in 1963. These changes led to a spate of low-scoring games for many seasons. This trend reached an extreme in 1968, when the league ERA and batting average, 2.98 and .231, respectively, were at their lowest since the 'dead-ball' era when game-used baseballs weren't replaced until they started to unravel. The next season the MLB made a drastic course correction, reinstating a smaller strike zone and lowering the height of the pitcher's mound from 15 inches to 10 inches. It's up for debate whether lowering the mound decreased or increased the rate with which pitchers suffer arm injuries, but it is certain that it had a positive effect for batters, rebounding to their historical averages.

While these tweaks to the physical playing field and strike zone have helped maintain the delicate balance of power between hitters and pitchers, an often-overlooked consequence is that these tweaks also change the accuracy of plate umpires. Lowering the pitcher's mound curbs the amount of downward momentum that pitchers can use, but in most cases the slightly reduced velocity will not help umpires. Moving the pitcher's mound back, even slightly, will force pitchers to work harder as well, but will have a larger consequence to the hitter and plate umpire's perceptual system by increasing the amount of time the ball is in flight and allowing for more information to enter the eye. Even an increase of a twentieth of a second of visual information could have massive implications in calling balls and strikes accurately. Other strategies such as growing or shrinking the strike zone may help batters exclusively while providing no real advantage to umpires. In fact, a smaller strike zone may make umpires less accurate, as the area of 'borderline' pitches begins to shrink. At the time of this writing, the MLB has proposed changes to shrink the strike zone in 2017, moving the lower boundary of the zone up several inches from the bottom of the knee to the top. It will be interesting to see whether this increases or decreases umpire accuracy now that we have detailed pre–post comparison data through the use of PITCHf/x.

The elephant at the ballpark remains the adoption of pitch-tracking technology to automatically call balls and strikes. Many fans remain deeply pessimistic of the technology and claim that the interchange between umpire and pitcher is a feature of the game, rather than a bug. Figuring out what a plate umpire's zone looks like is part of a pitcher's game preparation, rather than something to be ironed out and digitized. If a batter is called out with a borderline pitch, the batter's fans may call for an automated strike zone while the pitcher may be lauded for clever placement.

Some fans simply are against the intrusion of more technology into the game of baseball. These fans appreciate the consistency of the game, and that the critical equipment hasn't changed. The regulation baseball bat (no more than 2.61 inches around and 42 inches long, made of solid wood) and baseball (216 individual stitches around a cowhide cover, three layers of wool yarn, and a cushion cork pill) would feel as familiar to players today as to those from as far back as World War II. Many of the most lauded stadiums, such as Wrigley Field and Fenway Park, have a historical and cultural significance, while still being modern working sports venues. For many, changing the game of baseball is as distasteful as updating a Renoir or adding CGI characters into Star Wars. In recent years, the MLB has instated new rules to protect players, increase the pace of play, and increase accuracy through video challenges. Regardless of which new rules are proposed, there will always be a population of baseball purists that object to any change to the rules of the game.

Does baseball, or sport in general, become more enjoyable with the more accurate officiating? Transparently inaccurate play calling can severely damage enjoyment of a game, but not every injustice is created equal. When a home run ball is pushed foul by a gust of wind, fans understand that the game isn't always fair to players. We can accept that the game isn't equal or fair, and that is a major reason why we watch in the first place. If games didn't have element of unforeseeable provenience or tragedy, they wouldn't be worth watching. We want to see our team not just win, but overcome adversity. In this way, officiating doesn't need to be perfect. Officiating only needs to be accurate enough to make us feel as if the natural randomness of sport wasn't encroached upon by human bias or incompetence.

Fans should take a long look at whether features like PITCHf/x during live broadcasts increase the satisfaction they get from being spectators. We like knowing world records because we fantasize that maybe they'll be broken. We like knowing statistics because we think that maybe our team will perform better than average. This knowledge can get us more involved in the game, but I would argue that knowing the precise location of a pitch typically yields more frustration than enjoyment. Viewing a sport through a legalistic lens misinterprets why people watch in the first place. Fans do not choose optimal teams. Enjoyment isn't just about wins and losses—there are bad wins and good losses. Being a fan is more about fantasy, community, and entertainment than it is about the technicalities of the rules. Enjoyment requires a personal connection, and sometimes a sense of communal righteous indignation at the damn umpires is precisely what we need to connect with those around us.

From a psychological perspective, there is a very good reason to keep the human element in pitch calling—we like having someone to blame for our team's failures. Heckling is a tradition simply because when something bad happens humans need someone to blame. In order to preserve our state-of-mind, we often display a self-serving bias where we attribute success to ourselves and failure to outside circumstance (Beckman 1973; Miller and Ross 1975). When our favorite team dominates the opponent, we are more than likely to credit their superiority to the competition. When our favorite team crumbles, we begin searching for someone to blame, and umpires make for great scapegoats. While this bias clearly exposes our hypocritical nature, it also is in line with the unspoken rule all games share to simply 'have a good time'. While removing human error from officiating may make the game more precise, these biases, errors, and even blown calls are important elements in the hedonic enjoyment of a game. Hall-of-Famer Billy Evans, a major-league umpire from 1906 to 1927, was prescient when he claimed that 'The public wouldn't like the perfect umpire in every game. It would kill off baseball's greatest alibi—We wuz robbed'. It's just not as much fun to heckle a PITCHf/x camera.

Disclosure Statement

No potential conflict of interest was reported by the author.

Acknowledgments

The author would like to thank the 2016 World Champion Chicago Cubs.

References

ADAIR, R.K. 2002. *The physics of baseball: Third edition, revised, updated, and expanded*. 3rd ed. New York, NY: HarperCollins.
ASSOCIATED PRESS. 1939. Meter to record Feller's speed. *Richmond Times Dispatch*, June 6 Richmond, VA.

BASEBALL REFERENCE. N.D. Career Leaders and records for Batting Average. Available at http://www.baseball-Reference.com/Leaders/Batting_avg_career.shtml (Accessed 1 January 2016).

BALCETIS, E., and D. DUNNING. 2006. See what you want to see: Motivational influences on visual perception. *Journal of Personality and Social Psychology* 91 (4): 612–625.

BECKMAN, L. 1973. Teachers' and observers' perceptions of causality for a child's performance. *Journal of Educational Psychology* 65 (2): 198.

BHALLA, M., and D.R. PROFFITT. 1999. Visual–motor recalibration in geographical slant perception. *Journal of Experimental Psychology: Human Perception and Performance* 25 (4): 1076.

BORDNER, S.S. 2015. Call 'Em as they are: What's wrong with blown calls and what to do about them. *Journal of the Philosophy of Sport* 42 (1): 101–120. doi:10.1080/00948705.2014.911096.

BRUNER, J. 1957. On perceptual readiness. *Psychological Review* 64 (2): 123.

BRUNER, J. 1992. Another look at New Look 1. *American Psychologist* 47 (6): 780.

COLLINS, H. 2010. The philosophy of umpiring and the introduction of decision-aid technology. *Journal of the Philosophy of Sport* 37 (2): 135–146. doi:10.1080/00948705.2010.9714772.

DIMEO, N. 2007. Pitch f/x, the new technology that will change baseball analysis forever. *Slate Magazine*.

FANGRAPHS.COM. 2016. Major league leaderboards, 2015–2016, PITCHf/x data. Available at http://www.fangraphs.com/leaders.aspx (accessed 31 December 2016)

FAZIO, R.H., J.R. JACKSON, B.C. DUNTON, and C.J. WILLIAMS. 1995. Variability in automatic activation as an unobtrusive measure of racial attitudes: A bona fide pipeline? *Journal of Personality and Social Psychology* 69 (6): 1013.

GONZALES, M. 2015. Cubs manager Joe Maddon's umpire criticism gets MLB's attention. *Chicago Tribune*, 7 May. St. Louis, MO.

GOOLD, D. 2014. Baseball's new-age epidemic: Tommy John surgery, *St. Louis post-dispatch*, 25 May. St. Louis, MO. Available at http://www.stltoday.com/sports/baseball/professional/baseball-s-new-age-epidemic-tommy-john-surgery/article_aa8d6470-9f54-5851-af78-8afc725f9905.html

GREEN, E., and D.P. DANIELS. 2014. What does it take to call a strike? Three biases in umpire decision making. 2014 MIT Sloan Sports Analytics Conference, Boston, MA.

GREENWALD, A.G., and M.R. BANAJI. 1995. Implicit social cognition: Attitudes, self-esteem, and stereotypes. *Psychological Review* 102 (1): 4–27.

GREENWALD, A.G., D.E. MCGHEE, and J.L.K. SCHWARTZ. 1998. Measuring individual differences in implicit cognition: The implicit association test. *Journal of Personality and Social Psychology* 74 (6): 1464–1480.

KIM, J.W., and B.G. KING. 2014. Seeing stars: Matthew effects and status bias in major league baseball umpiring. *Management Science* 60 (11): 2619–2644.

KLAWANS, H.L. 1996. *Why Michael couldn't hit: And other tales of the neurology of sports*. New York, NY: Macmillan.

LIGHT, J.F. 2005. *The cultural encyclopedia of baseball*. Jefferson: McFarland & Company.

MAUS, G.W., J. WARD, R. NIJHAWAN, and D. WHITNEY. 2012. The perceived position of moving objects: Transcranial magnetic stimulation of area MT+ reduces the flash-lag effect. *Cerebral Cortex* 23 (1): 241–247. doi:10.1093/cercor/bhs021.

MCGOVERN, M. 1999. Toronto man has written the web site on heckling. *Reading Eagle*, 26 December, p. 29. Reading, PA.

MERTON, R.K. 1968. The Matthew effect in science: The reward and communication systems of science are considered. *Science* 159 (3810): 56–63. Available at http://science.sciencemag.org/content/159/3810/56.abstract

MILB.COM. N.D. Umpire Salaries. Available at http://www.milb.com/milb/info/umpires.jsp?mc=_ump_salaries (Accessed 1 January 2016)

MILLER, D.T., and M. ROSS. 1975. Self-serving biases in the attribution of causality: Fact or fiction? *Psychological Bulletin* 82 (2): 213–225.

MILLSLAGLE, D.G., B.B. HINES, and M.S. SMITH. 2013. Quiet eye gaze behavior of expert, and near-expert, baseball plate umpires. *Perceptual and Motor Skills* 116 (1): 69–77. doi:10.2466/24.22.27.PMS.116.1.69-77.

MOSKOWITZ, T., and L.J. WERTHEIM. 2011. *Scorecasting: The hidden influences behind how sports are played and games are won*. New York, NY: Crown Archetype.

NOSEK, B.A., M.R. BANAJI, and A.G. GREENWALD. 2002. Math = male, me = female, therefore math ≠ me. *Journal of Personality and Social Psychology*. Nosek, Brian A.: Dept of Psychology, U Virginia, 102 Gilmer

Hall, PO Box 400400, Charlottesville, VA, 22904–24400, nosek@virginia.edu: American Psychological Association. doi:10.1037/0022-3514.83.1.44

NOWAK, L.G., M.H.J. MUNK, P. GIRARD, and J. BULLIER. 1995. Visual latencies in areas V1 and V2 of the macaque monkey. *Visual Neuroscience* 12 (2): 371–384.

PARK, C.W., and LESSIG V. PARKER. 1981. Familiarity and its impact on consumer decision biases and heuristics. *Journal of Consumer Research* 8 (2): 223–230. Available at http://jcr.oxfordjournals.org/content/8/2/223.abstract

PAVLIDIS, H., and D. BROOKS. 2014. Framing and blocking pitches: A regressed, probabilistic model. *Baseball Prospectus*. Available at http://www.baseballprospectus.com/article.php?articleid=22934.

ROGERS, P. 1999. Umpire pay heavy dues on way to majors. *Chicago Tribune*, July 18 Chicago, IL.

RUSSELL, J.S. 1997. The concept of a call in baseball. *Journal of the Philosophy of Sport* 24 (1): 21–37. doi:10.1080/00948705.1997.9714537.

SHANK, M.D., and K.M. HAYWOOD. 1987. Eye movements while viewing a baseball pitch. *Perceptual and Motor Skills* 64 (3c): 1191–1197. doi:10.2466/pms.1987.64.3c.1191.

STROHMER, D.C., S.A. GRAND, and M.J. PURCELL. 1984. Attitudes toward persons with a disability: An examination of demographic factors, social context, and specific disability. *Rehabilitation Psychology* 29 (3): 131–145.

SUITS, B. 2014. *The grasshopper: Games, life and Utopia*. Peterborough, CA: Broadview Press.

THOMAS, A., A. DOYLE, and V. DALY. 2007. Implementation of a computer based Implicit Association Test as a measure of attitudes toward individuals with disabilities. *Journal of Rehabilitation* 73 (2): 3–14.

TOLLEY, M. N.D. The big list of umpire heckles. Available at http://www.heckledepot.com/umpire-heckles/ (Accessed 1 January 2016).

VICKERS, J.N., and R.M. ADOLPHE. 1997. Gaze behaviour during a ball tracking and aiming skill. *International Journal of Sports Vision* 4 (1): 18–27.

VOLKMANN, F.C. 1986. Human visual suppression. *Vision Research* 26 (9): 1401–1416. doi:10.1016/0042-6989(86)90164-1.

WITT, J.K., and D.R. PROFFITT. 2005. See the ball, hit the ball: Apparent ball size is correlated with batting average. *Psychological Science* 16 (12): 937–938.

WITT, J.K., S.A. LINKENAUGER, J.Z. BAKDASH, and D.R. PROFFITT. 2008. putting to a bigger hole: Golf performance relates to perceived size. *Psychonomic Bulletin & Review* 15 (3): 581–585.

YARROW, K., P. HAGGARD, R. HEAL, P. BROWN, and J.C. ROTHWELL. 2001. Illusory perceptions of space and time preserve cross-saccadic perceptual continuity. *Nature* 414 (6861): 302–305. doi:10.1038/35104551

INTERVIEW

Appendix: An Interview with Leonardo Fogassi*

Jens Erling Birch

Introduction (Rizzolatti and Fabbri-Destro 2010; Rizzolatti and Fogassi 2014; Rizzolatti and Sinigaglia 2008)

In the 1980s, a group of researchers led by Giacomo Rizzolatti in Parma, Italy was working on describing the organization of the brain's motor cortex. They were interested in finding out what was going on in the motor system of monkeys when they were presented with objects. The research group published an article in *Experimental Brain Research,* in 1988, presenting a new view on the organization of the ventral premotor cortex in macaque monkeys. (Rizzolatti et al. 1988). The article described how neuronal activation in area F5 correlated with specific goal-related motor acts rather than with single movements made by the animal. These neurons play a role in transforming visual information about an object into motor acts, and are now known as *canonical neurons*. The group kept on testing F5 neurons in the ventral premotor cortex by presenting food of different shapes and sizes, and surprisingly found that another set of neurons discharged when an experimenter grasped food. These neurons discharged both when the monkey performed a motor act and when it observed another agent performing a similar act. In 1991, the group sent a report to *Nature* about their discovery. The paper was not accepted, but was published *in Experimental Brain Research* the following year (Di Pellegrino et al. 1992). These neurons were later named *mirror neurons*. (Gallese et al. 1996; Rizzolatti et al. 1996).

After the discovery of mirror neurons, the question 'what is their function?' arose. Initially the group had two competing hypotheses: imitation and understanding. Imitation would seem to be the simplest explanation: transforming from vision to action through imitation. The problem with this hypothesis was that macaque monkeys are poor imitators and rarely imitate with their hands (like grasping). So, instead it was action understanding that seemed to be the most likely function of the mirror neurons. This was supported by experiments showing that neurons discharged in congruence with the goal of the motor act, but not with mechanical movements: similar movements with different goals did not activate the same neurons, but different movements with similar goals did. Understanding a motor act is an older evolutionary function, and imitation (and imitation learning) is more likely to have evolved on top of action understanding.

---

*An interview with Leonardo Fogassi. At his office at the Dept. of Neuroscience, Universita degli studi di Parma. October 10th 2016.

Although discovered 25 years ago, mirror neuron research is still in its infancy. What the practical implications might be are still debatable. How vision and motor action come together is important for both learning sport skills and improving them. If it is correct that vision and motor action are so deeply intertwined, then perhaps training and learning protocols should also reflect this insight. A more recent finding and interest is how mirror neurons contribute to our sense of space: mirror neurons play a role when it comes to distinguishing what is within an arm's reach (peripersonal space) and what is not (extrapersonal space). This sense of space plays a role in if/how we must move to grasp something. How near or far we must move to a ball or an opponent, say, must be of fundamental importance in sporting skills. Philosophically, the mirror neuron theory of action understanding is highly interesting when we consider and argue what skills are and how they are connected to the world. The mirror neuron system seems of great importance for high-level skills, especially in one-on-one sports, team sports, and sports where we respond to the actions of others immediately. Experts with high activation in the mirror neuron system are able to understand what another agent is doing faster. They discriminate more precisely, and they predict an outcome more accurately. The mirror neuron system helps us come up with efficient responses to opponents, and to pass the ball accurately to teammates. The theory gives a neural explanation of how it might be that we are able to respond quickly and effortlessly, seemingly without reflective thinking. This is also to say that we know something, without propositional content. The mirror neuron theory of action understanding also claims that there are intentions without reflective desires/beliefs. We are intentional, and thus conscious, even though we are not necessarily aware of the content of thought. The mirror neuron theory of action understanding points to the primacy of the motor system in both human evolution and daily life.

Interest in mirror neurons is, of course, not limited to sports. A more or less speculative hypothesis on a possible extrapolation of mirror neurons is their significance in the evolution of language. The idea is that language has not evolved primarily from the calls of animals, but gestures. The idea is that mirror neurons help us understand, for example, pointing gestures, and this again paves the way for attaching sounds to gestures. In addition, mirror neurons play a role in understanding mouth movements, which is related to the production of sound—and on top of that: imitational learning. These speculations are interesting, of course. On the one hand, they challenge some deeply rooted opinions about animal and human behavior and cognition. On the other, speculations and bold experiments might teach us something new. That is why research on autism and mirror neurons has received a lot of attention. Knowledge about autism is of course a good thing in itself, but it will also increase our understanding of the human mind more generally. If we find out why something does not work normally, we know more about what it takes for something to do so. For example: autistic agents might be able to understand the goal and intention underlying a motor act, but they do so differently than non-autistic agents. The latter understand the goal and the intention simultaneously; when you see someone picking up food (the goal), you already understand whether that someone will put it to the mouth to eat it (alternative intention #1), or place it somewhere else (alternative intention #2). Autistic agents understand the goal (picking up food) immediately, but understanding the intention is delayed until the movement to the mouth or somewhere else has begun. The same lack of understanding can be observed when it comes to understanding emotions: autistic agents may understand an emotion conceptually, but the feeling of an emotion is impaired due to a lack of activation in the mirror neuron system. The mirror mechanism of transforming observational information to personal knowledge might thus explain how we understand others' motor actions, emotions *and* feeling empathy (Damasio 2003). These are not only interesting from a philosophical point of view but also from a clinical one. Schizophrenic patients, for example, tend to show dysfunctional empathetic abilities and impaired understanding of others' intentions. That is why research on mirror neurons has also been linked to schizophrenia. The Parma group contributes with research on schizophrenia, especially on what constitutes the bodily self (Gallese and Ferri 2014).

Another application of mirror neuron research is related to neurorehabilitation: if you have a motor deficit and you want to move, then observing someone moving provides an extra neuronal spark, so to speak, by involving mirror neurons. Increased activation in the synapses may cause a positive loop, so that you may be up and walking sooner after a stroke (Buccino, Solodkin, and Small 2006). Mirror neurons are also said to play a part when we experience art, or have an esthetic experience (Gallese 2010) The idea is that our responses to art consist of the activation of embodied mechanisms encompassing simulation of actions and emotions. If so, then mirror neurons play a role when we are sport spectators as well. The first time that 'mirror neurons' really captured my interest was when I heard a presentation by Vittorio Gallese at a conference in Copenhagen, Denmark in 2006. Two years later, I spent several days at the Department of Neuroscience at the University in Parma. We were met with great enthusiasm by Giacomo Rizzolatti, spent a whole day talking and discussing with Corrado Sinigaglia, and were shown around the premises and laboratories the next day. In 2016, the same thing happened: I contacted Leonardo Fogassi and asked for an (short 30–40 min) interview. Instead, I talked to him for almost three hours, and then with Luca Bonini and Alessandro Livi for the rest of the day. I was shown new equipment for neural recordings and the room where they plan to film monkeys and record neural events while monkeys move more freely.

In the interview, I wanted to know more about the background for the discovery of mirror neurons and how research has developed from describing the organization of the motor cortex to recent findings, for example, how mirror neurons participate in coding space. I was also interested in how looking for visual properties in the brain's motor system, which was unorthodox in the 1980s, has undermined the view in cognitive science that perception, cognition and action are separate domains. The findings regarding the mirror neuron theory of action understanding show the significance of the motor system in other so-called higher cognitive functions, and I was interested in thoughts on the primacy of the motor system. Regarding the mirror neuron system's significance and relevance to sport, I wanted to know more about how the mirror neuron system contributes to fine-grained discrimination in action understanding. I was also interested in how/whether some persons have a better developed mirror neuron system enabling them to reach a higher level of sporting skill.

Professor Leonardo Fogassi is one of the researchers who have been present in the Parma group during these 30 years. He spent 15 months of his PhD in Richard Andersen's lab in the Department of Brain and Cognitive Sciences at MIT, and completed his PhD in neuroscience at the University of Parma in 1989. His research concerns the sensorimotor transformations operated by the cerebral cortex, and the role of motor cortex in action perception. These topics are studied by means of neurophysiological and behavioral techniques. Most of his publications are related to the neurophysiology of space perception, coding of prehension movements, and action understanding. In 2007, Fogassi was the co-recipient, with Giacomo

Rizzolatti and Vittorio Gallese, for the University of Louisville Grawemeyer Award for Psychology for their work on mirror neurons.[1]

The interview

Prof Leonardo Fogassi and I were talking about the weather, Norway versus Italy, skiing, and then after about half an hour we found ourselves talking about the Human Brain Project (HBP); a project initiated by the European Union, and regarded as being the neuroscientific counterpart to particle physics at CERN.[2] I was interested in why the HBP did not involve research on mirror neurons and why the Parma group did not participate in the project. Fogassi told me that, at least at the beginning, it wasn't possible to apply for new animal projects related to HBP, because HBP research is based on rodents, while the Parma group is expert in monkey experiments.

A: I don't believe that it is possible to directly make inference from mice to humans regarding many high order functions. I mean, the mice are very good as models, like mutant mice and so on of course. But in the case of high order capacities the distance is so far apart that it's very difficult to make hypotheses on the human brain starting from the mice. You need at least one primate intermediate. Indeed, there are people now trying to study mutant monkeys, to fill the gap.

Q: In Norway, we have May-Britt and Edvard Moser who received the Nobel Prize (together with John O'Keefe) in Physiology or Medicine in 2014. Their research on mice is on cells that constitute a positioning system in the brain while the mice are moving around.

A: You know, in that case it's ok…

Q: They don't talk about humans either, they just talk about the cells.

A: Yes, exactly.

Q: But I mean, your research here, if you could do single-cell recordings on mice or rodents in a maze, like they do, when moving; don't you think that would enhance some of your theories about movement and mirror neurons? If you could actually record neuronal activity while they were walking around?

A: In the mice you mean?

Q: Yes.

A: I know that…I mean, we are experts on monkeys, so to go to mice or rats would be…it's a kind of jump also technically, so it would take time. Actually there is a person at NTNU, I think it's Jonathan Whitlock who has an European Research Council grant exactly on this issue.[3]

Q: Ok.

A: So, it could be possible to find a way to demonstrate something about mirror neurons in the hippocampus.

Q: So, let's continue your criticism of the Human Brain Project from jumping to conclusions from rodents, mice, rats to humans. I guess you have the same problem when jumping from primates like macaque monkeys to humans. Doing research on single cell recordings on macaque monkeys and then trying to build theories, even cognitive theories, about humans and the mirror neuron system in humans, is *also* difficult.

A: It's difficult. But you know, primates are much closer to humans. And as you probably know well, there are a lot of homologies. The good thing is that with the monkeys you can also make neuroanatomical tracing experiments that can allow us to define the circuits, the networks, not only recording from single areas. You can describe a complete network, and then also record from areas in which you never recorded and so on. So, I understand the criticism, but on the other hand I think it's much more closer to…

Q: Of course, of course.

A: Also, this is actually one thing that we always write down in the proposals when we ask for permission, because in Italy, we have to ask for permission for experimental research on primates. What we point out is that the cognitive properties of monkeys are the only ones that could be studied with respect to the mouse's system, so we need to use primates because otherwise we could not address certain types of issues.

Q: It's the best you can have.

A: It's the best model that we have. Of course, one could use the apes, but it is not allowed due to the WMA Declaration of Helsinki.

Q: So, have you ever been close to having epileptic patients like Mukamel did, doing single cell-recordings on humans? (Mukamel et al. 2010).

A: The chance to record from single neurons? No. However, there is a good collaboration from our Rizzolatti group with Niguarda Ca' Granda hospital in Milan where they study epileptic patients. It is possible to implant electrodes for one or two weeks before surgery and this is a very good source of data. We were studying the possibility of using the same electrodes, also as microelectrodes, but it's quite tough. It's not an easy problem.

Q: But it's *possible* for your department to do direct recordings on humans? You have the possibility?

A: Well, of course you need to take a lot of steps, several steps in which they give you the permission to do it, and so forth. We have not arrived at this step yet. In principle, one could think about it. The problem, for example of the system used by Mukamel, Fried and colleagues, is that the microelectrodes they have are at the tip, so that they can record only from medial cortices, not from the surface, for example. Instead, one would like to have a system like we are now using in monkeys. If you speak with Luca Bonini, he can tell you some technical aspects of these electrodes in which you have a probe, and along the probes there are many contacts that allow us to record single neurons at different depths.

Q: Yes. So, let's go back to the start, in the 90s. Were you actually there?

A: Sure, we were a group.

Q: Can you tell us about it? When you read about it, it sounds like it was an Eureka moment—like some claim was the case when Watson & Crick discovered that the DNA molecule exists in the form of a three-dimensional double helix - that it just happened by chance, so could you try to describe it by your own words? (Watson and Crick 1953).

A: We were recording from premotor area 5 (F5) of the cerebral cortex, because earlier in this area we recorded neurons coding the goal of grasping acts. However, during these recordings we found that there was a certain percentage of neurons responding also to the simple presentation of objects. So, we were interested to see in a more controlled way the characteristics of these neurons. And another issue was - the link probably is interesting for you - the link between kinematics and grasping, and the neuron's response during grasping, to find out if there are some kinematic parameters that could be correlated to the neuronal discharge. We were interested in all these things. However, our approach was always a kind of, let's say clinical approach, in the sense that we were used to making a lot of different natural tasks in which the monkey could interact with us: we could present many types of visual stimuli, tactile stimuli and so on. This was a kind of approach introduced by Rizzolatti, who by the way at the beginning was a neurologist. So, the clinical approach is typical of a neurologist, but apart from this there is also a rationale, that is if you study an area just with a specific paradigm, at the end you can make conclusions based only on the variables studied in the paradigm; therefore, you could miss the function of a particular area. We are convinced that you can map the brain and you can find different areas that are involved in specific functions. We are also convinced also about the presence of borders that divide the areas and so on. So because of this we think that an area has a particular function and we were trying to investigate more of its properties. Using this natural approach we often move the hands and grasp things, give them to the monkey, maybe we go around in the lab, bring other things and so on. So, at a certain point we realized that there were neurons that responded not just when there was an object but when *we* approached the object with the hand. We were a bit surprised, but we went ahead with our idea. Then, however, we became more attentive to these kinds of things and we found that it happened again, and again. So, at a certain point after a few weeks we accumulated a series of reports on these neurons. At the time we had protocols in which we wrote in details what we did, such as the neuronal specific response, the stimuli or the behavior to which the neuron did not respond, just to have a control, and so on. In some cases we also made acquisition of controlled trials. So in the beginning it was just like an anecdotal finding. But then we noticed that there was a kind of relation between all these responses: we found the neurons responding when food was put into the mouth, some other neurons responded to breaking of food. So, there were two possibilities: either to describe each neuron, but then each neuron became an experiment, or try to find an explanation for all these types of neurons. And the explanation was already provided by our previous idea that in the motor system, when we find for example visual responses, these are not simple visual responses but they have already another form; they are in terms of motor representation. This was really the important thing, if we had not this idea before, we could not be in the position to interpret the discharge of these neurons in the way we did: not just as a perceptual response but as a response referred to the motor representation. This was really the concept that linked all these responses together.

Q: But you were working inside that framework, a paradigm with those sorts of theories, action/execution models?

A: Exactly. So, at a certain point we were thinking to write an article. We decided that maybe we could write down a note, because at that time it was typical to write down a research note before the paper in extenso. We decided to write a note because it seemed we could build a story on these neurons. At the beginning we sent it to *Nature* at, but they didn't accept it.

Q: They didn't accept it? That's very interesting. Why didn't they accept? What were their reasons for rejection? Were they interested in the topic?

A: They were not interested.

Q: That's funny.[4]

A: The research note was accepted in *Experimental Brain Research* (Di pellegrino et al. 1992). Then we published the full paper four years later in *Brain* (Gallese et al. 1996). But in parallel we were investigating two other issues that actually had the same interpretation because we were studying the pragmatic responses to objects, that were anyway visual responses; and the responses of neurons of premotor area F4 to stimuli introduced in the peripersonal space. We did all these papers almost at the same time.

Q: So already in the 90's you were doing experiments on extrapersonal and peripersonal space?

A: Yes. Actually, the first study done on this topic was in 1983 with a very natural approach (Gentilucci et al. 1983; Rizzolatti, Matelli and Pavesi 1983). Then we used a kind of robot that allowed us to control velocity, distance and so on. We could demonstrate in a more controllable way the presence of these types of neurons. This was published first as a research note in 1992 (Fogassi et al. 1992), and then in the extended version in 1996 (Fogassi et al. 1996), at the same time as mirror neurons. The idea is similar: we have a motor presentation of several types of motor acts and we map the visual responses on this type of representation. This allowed us to: in one case, understand the action of others, in the other case to understand objects or space.

Q: So, I'm wondering about understanding the intentions of others: how did you go from recording that the same neurons fired when a monkey observed some motor act and performing the same act, from that empirical observation to the theory of action-understanding? So how did you go about it? I guess you had different hypotheses. How did you try to find support for some and falsify others? What kind of experiments did you work out in the framework that you were working within?

A: We did a number of studies and two of them refer to the fact that these neurons recognize the aim of the motor act. For example, the study we did using the screen in which the experimenter was doing an action to disappear…

Q: Disappear, yes, like behind a wall.

A: Behind a screen and there was still a response when the monkey could not see what the experimenter could do—but knew of the action beforehand. While there wasn't when the same action was made *without any* object. This was important because it demonstrates that it isn't solely a visual response. A visual response to actions was already demonstrated in the temporal cortex by Perrett (Perrett, Rolls, and Caan 1982), but *this* is a real visual response. The output of these neurons are visual, while *in our case* the important thing is that the output is motor.

Q: It wasn't visual because the action was hidden behind a screen.

A: Exactly. The other important thing was that on audio-visual mirror neurons in which we found that there are neurons responding also to the sound of the motor act in the sense of…

Q: Was it the breaking of peanuts?

A: Like breaking peanuts or manipulating. All noisy motor acts that you can imagine. This demonstrates that if you have a motor representation, then this can be accessed by different types of inputs, in some way like language does (Kohler et al. 2002).

Q: In philosophy, when we talk about intentions, we usually talk about desire/belief, or reasons. Together with Rizzolatti you've recently published a review article where you state that it's difficult, of course, to talk about reasons like desires, beliefs - or propositional attitudes- and mirror neurons, but in philosophy it's sort of difficult to talk about intentions as something more and above the goal of an act without talking about desire/beliefs (Rizzolatti and Fogassi 2014). So, how do you go from understanding a *motor act* to interpreting another's *intention* without grasping the propositional attitudes on a cognitive and reflective level?

A: Well, one step further that we did in 2005 was the group of experiments where we studied *actions* instead of motor *acts* (Fogassi et al. 2005). In these experiments the rationale was that if we make a motor act, this motor act can be involved in different actions. If it is involved in different actions, the final goal of the action, that is the behavioral goal, not just the goal of the motor act, is different.

Q: So when you talk about the behavioral goal of grasping to eat, is that not to be hungry or just to put the food in the mouth?

A: To put the food in the mouth.

Q: Ok.

A: Of course, this is motivated by the fact that you need to eat. So we did these experiments in which we asked the monkey to make two actions: grasping to eat or grasping to place. We could demonstrate first of all that motor neurons, actually the parietal cortex first of all, which is not the classical motor region, were discharging differently during this type of grasping acts: 65% had a differential discharge. It was the same grasping motor act but the final goal was different. The placing was performed in two different ways, so that we could respond to the possible criticism that it was related to some kinematic aspects, and we found exactly the same result. Then we did the same task recording from mirror neurons in the parietal cortex, then we tried also in the premotor cortex and we found that actually a higher percentage of neurons could differentiate during grasping observation according to the different actions. Of course, the context was helping the monkey to differentiate between the two actions, otherwise it would not be possible. We found that during the observation of grasping there were the same responses and differential activity we found in the motor neurons. At the end, we compared the visual and motor responses of mirror neurons and found that if a neuron prefers grasping to eat during the execution, it prefers grasping to eat also during observation. So our idea is that in this case the discharge of the neuron during the motor task reflects what the intention of the agent is. The intention relates of course to the action's final goal. In the case of observation the differential discharge predicts what is the outcome of the action performed by the observed agent. Of course, this is a kind of basic circuit. We could say that both in the parietal and premotor cortex there is this type of property, although in the parietal it's more emphasized. We hypothesized that this is a kind of primitive basis for intention understanding.

Then, there were also studies on humans related to this issue (Iacoboni et al. 2005). I'm used to making a kind of difference between actions that are clear so that you can easily understand what the intention of the agent is, and actions that are ambiguous so that it is difficult, given the context, to understand the intention of the agent. In such cases you probably need a kind of inferential process that allows you to understand the reasons that are behind other's behavior. Certainly, in more ambiguous cases, if you make an fMRI study, you would have an activation of the mirror system, but probably you couldn't find a difference between the two actions because you need information for disentangling the two actions. So, I think that in such cases the mirror system needs collaboration with other areas that allow to make this type of disentangling. However, consider that in our normal life, very often we are exposed to actions that are clearly understandable. This is the reason why we call this automatic understanding. If you look at the responses of mirror neurons during the observation of actions, they are completely aligned with the action, not later, thus indicating that this response is not the result of an inferential process, that should take more time to occur.

Q: Ambiguous actions is an interesting topic I think, especially for motor skills, for example when you try to fool or trick an opponent in tennis or football. Have you tried to fool monkeys; Simulating that you are grasping, but your actual intention is something else? How does the mirror neuron system help us to see through such trickery?

A: The problem is that you need information. If information is given in a way that the monkey can understand, like a specific action, then the neuron will respond according to what the monkey understood. But you can fool a monkey of course.

Q: Easily. But what would your thoughts be concerning humans and ambiguous actions?

A: This I don't know experimentally. As an observer, you can detect specific features that allow you to recognize that there is an attempt to fool you, like kinematic features or expressions for example. This is an issue that must be completely explored. I don't know if there are people trying to explore it.

Q: Do you think it will be difficult in a laboratory setup with a monkey to try to fool it, to do different things, start a movement going this way and then going another way?

A: Yes, it is possible to do. One should prepare a context that points in one direction, and then make an action that is different. The problem is that, for example, in the task we used, only *after* grasping the monkey is able to understand what really could be the action. Based on this context the monkey's understanding is simply related to what the context allows it to infer.

Q: So understanding is not that early?

A: It's not very early. However, we had a study in which we made a modification of the task: it was run in two steps, so it was a kind of more complex task with respect to what I just mentioned (Bonini et al. 2011). There was a container with a cover, so the monkey had to grasp the cover and then cover the container, then it could either grasp the food and put into its mouth or grasp an object and place it into the container. However, there was a condition in which the monkey could see what the object put into the container was, before covering it, then there was a condition in which there was an opaque screen, so the monkey couldn't see and the first grasping was made in absence of any information of what the object could be. We found that when the monkey could see which object was put inside the container, the neuron could differentiate between two different actions also during the first grasping of the cover, because the monkey *knew* that there was a particular object inside. This demonstrated it is not just related to the vision of the object, but in this case also to the memory of it. When there was an opaque screen, the discharge of the neuron was the same in the two conditions, it was kind of half of the response obtained during the preferred conditions. Then, after grasping, the monkey could see the object and in the second grasping the neuron was capable of differentiating between the two actions. So, this means that when the monkey doesn't know which action is to be performed, instead of withholding the action, performs it but the neuron signals that there are two possible solutions. Only when there is additional information the monkey, the neuron actually, is able to differentiate.

Q: This brings us to the relation between mirror neurons and memory: in your example the monkey knows, or the neuron knows, what is in the container. On what level would you say that it knows? Is it still in the working memory? Is it in short-term memory, or is it already in the episodic long-term memory, having grown synaptic connections?

A: Let's remember that in these experiments between presentation and action there isn't a big delay.

Q: No. We're talking seconds.

A: To respond better to your question regarding working memory, we should elongate the distance, the duration of course. I would say that it's both because the monkey was primed to this task, so certainly it's in the long-term memory. On the other hand, I'm sure that if you change the type of object, the type of order, the monkey and the neuron are still able to perform this task. In this case you cannot speak about long-term memory of course. There is a problem of generalization, but I am convinced that if you think it in terms of areas, certainly the prefrontal cortex has an important role in this kind of task. Indeed, we recently demonstrated that using this task, there is activation not just during the grasp but also before (Bruni et al. 2015). So, probably the prefrontal cortex is organizing this type of actions, and it is interesting that the neuron responses prefer in general the object rather than the food: grasping to place is more represented than grasping to eat. Probably this is very important in relation to learning because grasping to place is learned, while grasping to eat is already in the hardware.

Q: It's like an evolutionary trait to eat.

A: Exactly. So, the idea of a further experiment would be to train the monkey to make let's say two, three learned actions and to see what are the differences are between the neuronal responses.

Q: This is very interesting. There have been a couple of studies on humans, fMRI studies, Calvo-Merino on dancers and Aglioti on basketball; there are more activations in the mirror neuron system in the brain of the expert dancers than the novices, and the experts are better at predicting basketball trajectories (Calvo-Merino et al. 2005; Aglioti et al. 2008). So, properties of mirror neurons must be an evolutionary trait in all primates, humans and probably other mammals as well. What do you think about epigenetic factors that are learned? It seems like experts have a better mirror neuron system at discriminating and understanding the actions of others.

A: I will respond to your question but before I do that, I would like to know your interpretation of this. You would say 'yes'?

Q: There are so many interesting issues related to this question. Let's say you have an evolutionary developed mirror neuron system; it must be a genetic feature of the species, enabling you to respond faster, and hence increase the possibility for survival. That should be the evolutionary benefit by having a mirror neuron system on a species level. Still, you can develop the mirror neuron system further to an expert level on much more fine-grained actions it seems, like dancing and basketball - which are not evolutionary movements. The movements are enabled by our species' anatomy, but we're not evolutionarily made to play basketball obviously. It's understandable, but still it's not part of an evolutionary system. So how do you get a better mirror neuron *skill*, if you can call it that, a skill of discriminating actions, understanding intentions on a micro level, understand in soccer or tennis that you are trying to fool me? Obviously experiential learning. The better players probably learn facial expressions and bodily movements that they're exposed to. Synaptic growth in special areas of the brain must be established in order to detect; they are learned in that way.

A: However, the learning is based on a network that is already there.

Q: Of course. Do you think there are individual differences - are there people who have a better genetic make-up to evolve the mirror neuron systems?

A: Let me take an example. My colleague and myself (actually we started together, then my colleague continued on the topic) performed a study on neonatal imitation in monkeys - neonatal imitation is limited to a brief period in monkeys, seven days let's say (Ferrari et al. 2006). Then in a further study he found that, as actually was demonstrated in humans many years ago (Meltzoff and Moore 1977), if you take 100 monkeys, and see that 50 show neonatal imitation and 50 do not show it, if you make a follow-up of the same monkeys three years later on several traits, you find that those that showed neonatal imitation were better in motor skills and also in social interactions. So, this reveals that probably there is a difference already at birth. It's something that probably is needed because it's in a *critical* period, likely to create an immediate bond between the mother and the newborn, and so on.

Q: Do you think there are also *sensitive* periods for developing a mirror neuron system in humans, to get better at discriminating and understanding actions?

A: Probably yes, but I think there is no study that really demonstrates that. Coming back to your question about the dancers and ballplayers: our interpretation is that there is plasticity in the motor system. So, first of all there is a structural change, a modification in the circuit that is related to these types of motor skills, on which these motor skills can rely. Then, after these plastic changes occurred, observations reveal that you are more of an expert. Probably during training you observe and then imitate and so on, but the starting point is always the motor system. People who don't know too much about mirror neurons believe that it is the perceptual system that is in some way plastically modified. However, these perceptual studies simply reveal what's happening in the motor system.

Q: So you believe that the motor system in the brain is primary to the visual system?

A: Absolutely.

Q: This is one of the most important philosophical and psychological consequences of your theory, I believe: that it seems to be a move away from information processing theories, going from perceptual inputs, to cognitive decisions and then a motor output.

A: Like a hierarchical system.

Q: Right. There are also more and more discoveries of mirror neurons in different brain regions, and there will probably be more and more discoveries that neurons have several properties, not just one—which I also believe is one of your most interesting empirical findings. Do you feel that the standard cognitive theory of information processing are vanishing, or are people still holding onto them and criticizing your proposals?

A: I think both.

Q: Is there a change?

A: I think the idea you just described about information processing is still very strong. Simply because artificial intelligence was built strongly on this type of interpretation. We just compared this interpretation in a work in which we compared the different perspectives from which one can see an action (Caggiano et al. 2011). We found that among 100 neurons, let's say 26% are invariant with respect to the perspective. The other 74% are variant and are equally subdivided between different perspectives: egocentric, lateral and frontal. This is very important, because the output of a neuron is always only one …

Q: Yes, like on or off.

A: Well, the fact is that the output that comes out from the neuron, from the axon, that is only one. The dendritic three is very wide, so the neuron can receive a lot of inputs, but the output is only one, that is only one message. It can be modulated, but it's only one message. So, what is the message? If you are in the motor system, the message is motor. So, how can a neuron which tells you what is the goal, have the possibility also to tell you what is the perspective from which this act is made, or what is the value of what you are grasping for? In comparison with the hierarchical interpretation, our interpretation is that in the temporal cortex - the action observation circuit is temporo-parieto-frontal –there are these neurons that respond to the visual features of the action, and also to the different perspectives. Our idea is that when you see an action made from a particular perspective there is an activation of the circuit in the temporal-to- frontal direction, but then the output of the mirror neuron that receives this input is in the motor cortex and reveals what is the goal - but at the same time comes back to the temporal cortex and emphasizes the perspective that the monkey is actually observing from. That is a kind of reciprocal connection that allows you to understand *both* the goal and the visual perspective of an action.

Q: What you are describing sounds a lot like Edelman's theory of what he calls the reentrant loop (Edelman and Tononi 2000).

A: In some way, although there are also dedicated circuits, very dedicated. On the other hand, let's take the example of motor imagery: I ask you to imagine to grasp. In this case, in terms of timing, the first activation you have is in the motor cortex. Our idea is that your motor repertoire is already present at birth, and of course during your life becomes modified, is refined and so on. But this motor repertoire constitutes your first knowledge, so that the output of the motor cortex is sent back to, for example, the parietal cortex where there are many neurons that then will receive information from the visual-, the acoustic-, and from the somatosensory cortex. These kind of inputs will serve as confirmation of your motor knowledge. So, if a specific grip means grasping with force, then the shape of the object will confirm this motor knowledge.

Q: It's a quite recent finding that mirror neurons also have inhibitory properties? (Vigneswaran et al. 2013).

A: You mean the Kraskov's findings?

Q: I think there are several themes that are interesting here: one is just concept 'mirror neuron'. If something is actually mirrored, being strongly congruential or identical neuronally, then I should imitate automatically what you do, which I don't. There must be some kind of inhibitory mechanism so I just don't do exactly what you do, and there is! So, do you think now, after 20 years, that the concept 'mirror neuron' is a good one since activation in doer and observer are not actually identical? Or do the recent findings of inhibitory properties of mirror neurons support the function of action understanding on an even more fine grained level?

A: Well, I think that the concept of mirror neuron is still good because it explains well what is an automatic form of understanding. It can be applied to monkeys, it can be applied to humans, and it can be applied to singing birds. It's a mechanism that you can apply to the motor system, but also to the emotional system. You can also apply it to peripersonal space, so there has been a widening of the concept of the mirror neuron system, but all these aspects refer to the same neural mechanism. It is important to have found inhibition in mirror neurons because it reveals that *there is* an inhibition. Earlier this was only hypothesized.

Q: Why did it take so long to find inhibition in the mechanism?

A: Well, because actually it is difficult to devise an experiment that can allow you to see an inhibition because when you are observing something, there is always a motor inhibition, so you should be so lucky to find a neuron that is involved in this inhibitory mechanism. I'm not sure that the mechanism, the neurons found by Kraskov, Lemon and collagues, are the only neurons that have these inhibitions, but their work at least reveals that during observation there is also an inhibition. Their study was interesting because they did also find mirror neurons that have an output to the spinal cord, so in principle that could also have an effect on movement and on blocking the motor responses similar to the observed ones. On the other hand, you know that there are these people who tend to imitate everything you show (echopraxic patients).

Q: Do you think that's a deficit in the mirror neuron system?

A: Probably, yes.

Q: What about other hypothesis' about what mirror neurons do? Ramachandran talks about autism, others about language learning, social and emotional responses, mindreading and so on. How do you feel as a scientist about all these things? Is it just bad science?[5]

A: As we discussed before, such functions must be explained by the work of mirror neuron system together with other areas. For example, also imitation which seems so well suited for the mirror system requires the intervention of other areas. Observation is something passive, while when you imitate you need an intervention, a voluntary decision, so you need other areas as well. About language learning; I am convinced that is a very important aspect where mirror neurons are certainly at work during the first years of development, but also in adults.

Q: Do you think it's both movement of the mouth and the sound that trigger the mirror neuron mechanism?

A: Well, together. There are probably neurons in the ventral premotor cortex that control both, so here I think the theory of mirror neurons provide a very good model. When it comes to emotional reactions, I think that the mirror neuron system is important for understanding, but for a reaction you need also other areas. Otherwise you would see an emotion and immediately produce this emotion. So, you need an inhibition again, but you need also a mechanism that allows you to react.

Q: So again; do you think the motor system in social and emotional contexts is primary to, so to speak, more social learned responses?

A: Yes.

Q: So even here you start with motor system?

A: Even in emotional understanding. It is true that you have an involvement of areas that are not primarily motor, but if you look at their output, their output is viscero-motor. So the principle is the same.

Q: Would you go as far as claiming that our human motor capabilities or motor systems are more fundamental to us than other so-called higher cognitive abilities?

A: Our idea is that the motor system is primary, and from this many cognitive properties have been built. When a new function has emerged, it is possible that you can build other circuits on these functions. I'm certainly not a follower of the idea that you build knowledge using your sensory systems. If you think about it, the motor system is primary also in unicellular organisms. The motor system is fundamental for our interaction with the environment, and to receive sensory feedback from it.

Q: Would you say that without the kind of anatomical make-up and the motor systems that we have, we would not have language, or invent planes and cars?

A: Language I think would be very difficult to have. You could make probably some sensory maps of the world, but without movement?! What can we explore, what can we change in the environment?

Q: It's like Aristotle said; locomotion comes before cognition.

A: Exactly, many philosophers have proposed a similar concept.

Q: But there are even more saying cognition is the most defining property of humans. Let us finish off with a couple of questions about the future. Now we're in 2016 - where do you see your department here in Parma going? What are your main research aims in the next 5, 10 years? That's one question. Two: if you could have your dream setup, what kind of laboratory study would that be, to accomplish all your goals and visions? If you could have everything you wanted?

A: Hahaha! The first question is really related to funds, so it is difficult to make any considerations without them. During the last decades there has been an enlargement in this institute of studies on humans and children. I think things like studies on the human brain directly with electrodes on epileptic patients, or on specific syndromes that can be related to the mirror system like autism or rehabilitation could be fields that can certainly be emphasized over the next years. Another possible field is that of schizophrenic patients. Vittorio Gallese is the person who is studying this, in particular related to hallucination and the bodily self (Gallese and Ferri 2014).

A: Returning to the question of studies I would like to do, there is one starting now ('Motor and cognitive functions of the monkey premotor cortex during free social interactions'). It has received an ERC grant and is led by my collaborator Luca Bonini and colleagues. They are starting a study in which the monkeys will be free, so it will be possible to really study the neuronal behavior of monkeys that are not sitting in a chair. You can then compare the experiments made in the chair with the experiments made in this bigger cage where the monkey will actually do the same thing but in a completely free situation. I will be very, very interested to see the results. It has been one of my dreams for many years, but it was not possible technically. Now, with telemetric recording it finally is. Another thing that would be most interesting is to demonstrate what are the bio-molecular aspects related to mirror neurons. We don't know anything about it. We would need either monkeys that are performing particular tasks related specifically to mirror neurons and not to other types of neurons, or we would need mutant monkeys that have a deficit in specific things related to social interactions. Finally, now that the neural network is more enlarged, mirror neurons have been found in different areas, and probably having different roles, it would be nice to see if these different roles are linked to specific deficits. Concerning this latter point, another thing that has been difficult to demonstrate is the knockout of the mirror system: for example, to make lesions or inactivations that allow you to see if the monkey is not able anymore to understand or to make the same types of behavior. This is very difficult, probably because there are many nodes in a neural network: if you inactivate one, its function can be partially substituted by another one. We actually tried this experiment many years ago, but it was not straightforward.

Q: Thank you, professor Fogassi for this long and very interesting talk.

Notes

1. See https://grawemeyer.org/2007-giacomo-rizzolatti-vittorio-gallese-and-leonardo-fogassi/
2. See https://www.humanbrainproject.eu/
3. See https://www.ntnu.edu/kavli/research/whitlock: The research aims to to describe the population coding of goal-directed movement intentions in the parietal and frontal areas of neocortex. By describing how these circuits work, the hope is to shed light on how we understand the actions of others.
4. Between 2012 and 2016, *Nature* and *Nature Reviews Neuroscience* published 11 articles on mirror neurons. https://www.nature.com/search?date_range=2012-2017&journal=nature%2Cnrn&order=date_desc&q=%22mirror%20neurons%22
5. See e.g. https://www.wired.com/2013/12/a-calm-look-at-the-most-hyped-concept-in-neuroscience-mirror-neurons/. For a solid, scientific review, see Kilner and Lemon 2013.

Acknowledgments

The author wishes to thank Prof. Fogassi for the opportunity to arrange the interview, and for taking the time to comment upon and edit the final manuscript.

Disclosure statement

No potential conflict of interest was reported by the author.

References

AGLIOTI, S., P. CESARI, M. ROMANI, and C. URGESI. 2008. Action anticipation and motor resonance in elite basketball players. *Nature Neuroscience* 11 (9): 1109–1116.

BONINI, L., F. SERVENTI, L. SIMONE, S. ROZZI, P. FERRARI, and L. FOGASSI. 2011. Grasping neurons of monkey parietal and premotor cortices encode action goals at distinct levels of abstraction during complex action sequences. *Journal of Neuroscience* 31: 5876–5886.

BRUNI, S., V. GIORGETTI, L. BONINI, and L. FOGASSI. 2015. Processing and integration of contextual information in monkey ventrolateral prefrontal neurons during selection and execution of goal-directed manipulative actions. *Journal of Neuroscience* 35 (34): 11877–11890.

BUCCINO, G., A. SOLODKIN, and S. SMALL. 2006. Functions of the Mirror Neuron System: Implications for Neurorehabilitation. *Cognitive and behavioral neurology* 19 (1): 55–63.

CAGGIANO, V., L. FOGASSI, G. RIZZOLATTI, J. POMPER, P. THIER, M. GIESE, and A. CASILE. 2011. View-based encoding of actions in mirror neurons of area F5 in macaque premotor cortex. *Current Biology* 21: 144–148.

CALVO-MERINO, B., D. GLASER, J. GRÈZES, R. PASSINGHAM, and P. HAGGARD. 2005. Action observation and acquired motor skills: An fMRI study with expert dancers. *Cerebral Cortex* 15 (8): 1243–1249.

DAMASIO, A. 2003. *Looking for Spinoza. Joy, sorrow and the feeling brain*. New York, NY: Harcourt.

EDELMAN, G., and A. TONONI. 2000. *A universe of consciousness*. New York, NY: Basic Books.

FERRARI, P., E. VISALBERGHI, A. PAUKNER, L. FOGASSI, A. RIGGIERO, and S. SUOMI. 2006. Neonatal imitation in rhesus macaques. *PLoS Biology*. 4 (9): 1501–1508.

FOGASSI, L., V. GALLESE, G. DI PELLEGRINO, L. FADIGA, M. GENTILUCCI, G. LUPPINO, M. MATELLI, A. PEDOTTI, and G. RIZZOLATTI. 1992. Space coding by premotor cortex. *Experimental Brain Research* 89: 686–690.

FOGASSI, L., V. GALLESE, L. FADIGA, G. LUPPINO, M. MATELLI, and G. RIZZOLATTI. 1996. Coding of peripersonal space in inferior premotor cortex (area F4). *Journal of Neurophysiology* 76: 141–157.

FOGASSI, L., P. FERRARI, B. GESIERICH, S. ROZZI, F. CHERSI, and G. RIZZOLATTI. 2005. Parietal lobe: From action organization to intention understanding. *Science* 308 (5722): 662–667.

GALLESE, V. 2010. Mirror Neurons and Art. In *Art and the Senses*., edited by F. Bacci and D. Melcher. Oxford: Oxford University Press: 441–449.

GALLESE, V., and F. FERRI. 2014. Psychopathology of the bodily self and the brain: The case of schizophrenia. *Psychopathology* 47 (6): 357–364.

GALLESE, V., L. FADIGA, L. FOGASSI, and G. RIZZOLATTI. 1996. Action recognition in the premotor cortex. *Brain* 119 (2): 593–609.

GENTILUCCI, M., C. SCANDOLARA, I. PIGAREV, and G. RIZZOLATTI. 1983. Visual responses in the postarcuate cortex (area 6) of the monkey that are independent of eye position. *Experimental Brain Research* 50 (2-3): 464–468.

IACOBONI, M., I. MOLNAR-SZAKACS, V. GALLESE, G. BUCCINO, J. MAZZIOTTA, and G. RIZZOLATTI. 2005. Grasping the intentions of others with one's own mirror neuron system. *PLoS Biology* 3 (3): 529–535.

KILNER, J., and R.N. LEMON. 2013. What we know currently about mirror neurons. *Current Biology* 23: R1057–R1062.

KOHLER, E., C. KEYSERS, M. UMILTÀ, L. FOGASSI, V. GALLESE, and G. RIZZOLATTI. 2002. Hearing sounds, understanding actions: Action representation in mirror neurons. *Science* 297: 846–848.

MELTZOFF, A., and M. MOORE. 1977. Imitation of facial and manual gestures by human neonates. *Science* 198: 75–78.

MUKAMEL, R., A. EKSTROM, J. KAPLAN, M. IACOBONI, and I. FRIED. 2010. Single-neuron responses in humans during execution and observation of actions. *Current Biology* 20 (8): 750–756.

DI PELLEGRINO, G., L. FADIGA, L. FOGASSI, V. GALLESE, and G. RIZZOLATTI. 1992. Understanding motor events: A neurophysiological study. *Experimental Brain Research* 91: 176–180.

PERRETT, D., E. ROLLS, and W. CAAN. 1982. Visual neurones responsive to faces in the monkey temporal cortex. *Experimental Brain Research* 47: 329–342.

RIZZOLATTI, G., and M. FABBRI-DESTRO. 2010. Mirror neurons: From discovery to autism. *Experimental Brain Research* 200: 223–237.

RIZZOLATTI, G., and L. FOGASSI. 2014. The mirror mechanism: recent findings and perspectives. *Philosophical Transactions of the Royal Society of London, series B Biological Sciences*, 369 (1644): 1–12.

RIZZOLATTI, G., and C. SINIGAGLIA. 2008. *Mirrors in the Brain*. Oxford: Oxford University Press.

RIZZOLATTI, G., M. MATELLI, and G. PAVESI. 1983. Deficits in attention and movement following the removal of postarcuate (area 6) and prearcuate (area 8) cortex in macaque monkeys. *Brain* 106: 655–673.

RIZZOLATTI, G., R. CAMARDA, L. FOGASSI, M. GENTILUCCI, G. LUPPINO, and M. MATELLI. 1988. Functional organization of inferior area 6 in the macaque monkey. *Experimental Brain Research* 71: 491–507.

RIZZOLATTI, G., L. FADIGA, V. GALLESE, and L. FOGASSI. 1996. Premotor cortex and the recognition of motor actions. *Cognitive Brain Research* 3: 131–141.

VIGNESWARAN, G., R. PHILIPP, R. N. LEMON, and A. KRASKOV. 2013. M1 corticospinal mirror neurons and their role in movement suppression during action observation. *Current Biology* 23: 236–243.

WATSON, J. D., and F. CRICK. 1953. Molecular structure of nucleic acids: A structure for deoxyribose nucleic acid. *Nature* 171: 737–738.

Index

Note: Page numbers followed by 'n' refers to endnotes respectively.

Abi-Rached, J. 27, 29
The Absent Body (Leder) 36
academic argument 7
action understanding 83, 84–9, 91, 92, 94, 138–40, 149
a fortiori 44
Agassi, A. 41
AI *see* athletic identity (AI)
alcohol 9, 25, 31, 108, 109
Ali, Mohammed 115
Alpine skiing 65, 67, 68, 70, 73, 75, 76
Alzheimer's disease 25, 26, 39
amateur boxing 16
ambiguous and arbitrary 15
American athletes 105
American Football 3, 7, 8, 12–14, 16–19
American Medical Association 16
amphetamines 55, 108
anticipatory discourse 28–9
anti-doping policy 58
anxiety 3, 42, 72–4, 119
aporia 68
archery 57, 119
Arendt, H. 27
argument from exploitation 7
argument from fraud 7
Aristotle 14, 27
athletes 27, 51, 92; chronic traumatic encephalopathy in 26; elite 49, 50, 59; performance of 52
athletic brain 73
The Athletic Brain: How Neuroscience is Revolutionising Sport and Can Help You Perform Better (Katwala) 73
athletic identity (AI) 49–51, 59
athletic personality and technological 50–2
automatic understanding 86
Autonomic Nervous System (ANS) 73
autonomy 8–15

autonomy-limiting activities 15
autonomy-limiting decision 11

banning of boxing and mixed martial arts 16
Bar-Ilan, O. 33n1
baseball 92, 105, 107, 108, 127–35
BASE jumping 15
basic minds and human cognition 118
Beck, Ulrich 28
Being and Nothingness 66
Belcher case 31
benevolence, principle of 12
Bentham, J. 13
Bergman, I. 75
Bernard, C. 70
biopolitical shift 27
Birch, J. 3, 4, 90, 93–4
The Body Keeps the Score: Brain, Mind, and the Body in the Healing of Trauma 40
bodyminded states 113, 119, 124n1
boxing: debates in 6–7; head in 8–9
brain-body dualism 37
brain imaging 26, 29–32, 84, 87, 95n8
Brain in flight: The anxious brain in action (Roelofs) 72
brain injury 3, 38–41; in football 7–8; risk of 6–7, 10–11
brain, physiological interventions 56
brain stimulation techniques 56–7
Brainwashed 30
Brave New World (Huxley) 109
Breivik, G. 68, 90–4
bridge-crossing-person 15
Brink, D. O. 14
British Medical Association 16
The British Medical Journal 16
broadly congruent mirror neuron 83
Broca, P. 38–9

INDEX

Brooks, R. 121
bungee jumping 15
Bush, George H.W. 1

caffeine 56
The Call of the Wild (London) 71
Calvo-Merino, B. 92
canonical neurons 82, 83, 138
Carr, D. 104
Carrio, A. 61n7
Cartesian dualist intelligibly 38
catchers 108, 130, 131
cause-and-effect relationship 17
Chalked Up: Inside Elite Gymnastics' Merciless Coaching, Overzealous Parents, Eating Disorders, and Elusive Olympic Dreams (Sey) 41
champion 50, 54
chemical enhancement 108–9
chemical intervention 108
Chicago Cubs players 132
chronic traumatic encephalopathy (CTE) 16–19; anticipatory discourses 28–9; concussion and 25–6; diagnosis 30
Churchland, P. S. 102
cingulate cortex 82
Clarapede, E. 40
climbing 122
cocaine 56
cognition process 4, 53, 113, 118; motor actions 89
cognitive abilities 50, 52–4
cognitive emptiness 122
cognitive enhancement, criticisms of 57–9
cognitive enhancers 54–6
cognitive functioning deficits 26
cognitive improvement 53
cognitive knowledge 92
communitarian approach 104–5, 109, 110
conception of vulnerability 17
The Concept of Mind (Ryle) 100
concussion 25–6; imaging technologies 29
Concussion in Sport Group (CISC) 2
Concussion in Sport Group (CISG) 18
conditioned and unconditioned reflexes 73
conscious experience 37
conscious feelings 37
consensual domination 11–12, 15
conservative approach 108
coordination 101, 102
Corlett, J. Angelo 7–8, 19n2
Cozolino, L. 37, 39, 41
credibility 29
criminalization 12
cross-cultural analysis 105
Crossing the Line: Violence and Sexual Assault in Canada's National Sport (Robinson) 41

Crosson, S. 68
Csikszentmihalyi, M. 106, 119
CTE *see* chronic traumatic encephalopathy (CTE)
culture of risk 27
cytoarchitectonic structure 84

Davis, N. J. 54, 56, 57
de Beauvoir, Simone 66
'decade of the brain' 24
decision-making 13, 14, 19, 28, 41, 43, 92, 129, 131, 132
DeLong, Jordan 4
Dementia 26
Dennett, D. 116
depression 26, 31, 42, 51
Descartes, R. 38
diagnosis 29–31
Die Kehre 66
direct mechanical trauma 17
discourses of risk and anticipation 28–9
Ditka, M. 44
Dixon, N. 8–11; Mills slavery analogy 11–18
dominant theory 17
dopamine 55
doping effect 58
Dreyfus, H. 90–2, 94, 116, 120, 121, 122, 124
Dumit, Joseph 30
Dworkin, G. 8, 11
Dylan, B. 44

earned run average (ERA) 128
economic argument 7
Edelman, G. 83, 86
electroencephalography (EEG) 84, 95n7
embodied cognition 114
emotion 3, 37–45, 51, 52, 54, 60, 74, 88, 90, 93, 94, 101, 106, 122, 124, 139, 140, 149, 150
emotional arousal affects memory processing 40
emotional pathology 42
emotional trauma and sport 40–3
enactive approaches 113
enactively embodied approach 123–4
enactivism 113, 120–3, 123–4; embodied cognition 114; human cognition and basic minds 118; mental representation 115–18
energy drink 10
Esposito, R. 27
ethical judgment 103
ethics of neuroscience 2
ethics, team spirit 102–3
Evert, C. 112, 116
Experimental Brain Research 138
extended mind 93
eye movement 129–31

INDEX

Fainaru, S. 16, 39
Fainaru-Wada, M. 16, 39
Fallgeschichte 69–70, 76
Farah, M. J. 1
Faulkner, A. 32
Fehlleistung 74
Feinberg, J. 8, 9, 12, 37–40
Felicific Calculus (FC) 13
Fiala, Andy 3–4
Field, Wrigley 134
fight–flight–freeze response 72
flow 106–7
Foddy, B. 53
Fogassi, L. 4, 86, 95n3, 138–51
football, elimination of 7–9
football players 9
Force Majeur 68
fore and enactivism 121
Foucault, M. 27
Frazier, Joe 115
Freud, S. 74
Fry, Jeffrey 3, 94
functional magnetic resonance imaging (fMRI) 26, 29, 30, 39, 83, 84, 87, 91–2, 95n4
future autonomy 10–11

Gammage, K. L. 102
Giddens, Anthony 28
Glannon, W. 103
Goldman, A. 89–90, 93, 94
good chemistry 101
goods argument 12–14
group flow 107
gymnastics 41

Hardes, Jennifer 3
hard paternalism, soft paternalism *vs.* 9
hard-wired fright response 73
harm and injury 37–8
harm principle 8
Harm to Others (Feinberg) 37–8
head in boxing 8–9
health care 32; and medical costs to others argument 7
Hebbian principle of learning 40
heckling in baseball 127, 135
Hegelian social ontology 104, 107
Heideggerian term 65–6, 75
Heidegger, Martin 66, 76
Hickok, G. 85–8, 90, 92, 95n10
high-level cognition 113
Hoberman, J. 49, 52
Honnold, A. 122
Hopsicker, P. 90–4
Howe, L. A. 77
human and monkey, mirror neurons in 81, 83–7, 130, 138, 140–51

Human Brain Project (HBP) 141
human cognition and basic minds 118
human consciousness 24
Hume, D. 13
Hutto, D. 114, 116–18, 121, 124, 124n2
Huxley, Aldous 109

Illes, J. 33n1
Ilundáin-Agurruza, J. 4, 94
imaging technologies 26, 29–32
implicit emotional memory 40
inasmuch 55
inattention and reasoning 26
inauthentic personalities 58
incompatibility 51
individualist 109
information processing theory 91
injury 37–8
inter-collegiate football, elimination of 7–8
internal representation 115, 116, 118, 121
Internet Movie Database (IMDb) 68
intervention 52
intrapersonal neutrality 13
involuntary harm 9
Isle of Man TT motorcycle racing 8

Jackson, Phil 102
Jensen, Frances E. 42
John, Tommy 129
Jordan, Michael 133
judgment and bias 131–3

Kahane, G. 58
Kant, I. 12, 104, 105
Kapur, N. 43
Katwala, Amit 73
Keenan, F. 68
Kelley, Joe 129
kendō 119
Khunian paradigm shift 82
Kim, Jerry 132
King, Braden 132
knowledge production 30
Krein, K. 4, 125n8
kyūdō archery 119

Lacan, J. 70, 76
law and neuroscience 28–9
lawsuits 23–4
Leach, J. 72
League of Denial: The NFL, Concussions, and the Battle for the Truth 39
Leder, Drew 36
LeDoux, J. 37
legal system 10
Levinas, E. 94
Levi, Neil 2

liberal individualism 103–6
liberal–libertarian tradition 104
liberals and libertarians 105
libertarian individualists 109
Lilienfeld, S. 30
limbic resonance 101
limits of team spirit 103–6
linguistic communication 101
litigation 3, 24, 26, 27
Little Girls in Pretty Boxes: The Making and Breaking of Elite Gymnasts and Figure Skaters (Ryan) 41
Loland, S. 51–2, 58–60, 62n34
London, Jack 71
Lopez Frias, Francisco Javier 2–3
Lovett, F. 15
Lupton, D. 30

macaque monkeys, mirror neurons 81, 83–5, 87, 130, 138
McNamee, Mike 2–3, 49, 58
Maddon, Joe 132
magnetic resonance imaging (MRI) 29, 39
magnification and staining techniques 39
Major League Baseball (MLB) 55, 127–9, 132–4
Malabou, C. 37, 39, 43–4
materialities 29, 30
Matthew Effect 132
medical diagnostics 31
medical epistemic uncertainty 16–17
medicalization of concussion knowledge 30
memory 40
memory loss 26
memory system 40
mental flexibility 40
mental representation 115–18
mental states and motor knowledge 90
methylphenidate and cognitive enhancement 55
mild traumatic brain injury (MTBI) 17, 36
Mild Traumatic Brain Injury Committee (MTBI Committee) 26
Millian argument 8–9
Millikan, R. 117
Mill, John Stuart 6, 8–9; ambiguous and arbitrary 15; to autonomy principle 10–11; consensual domination 11–12; goods argument 12–14; prohibition exists 16–18
mind–body dualism 38
mind–body relationship 38
mind does 89
mirror neuron system (MNS) 81–3, 92, 138–41, 144–51; efficient 91; function of 86; Hickok, Gregory 85–8; in humans 84–5; nurturing 91–3; philosophical consequences 88–90; philosophy of sport 90–3; trainability of 92
MLB *see* Major League Baseball (MLB)
MNS *see* mirror neuron system (MNS)

modafinil 55
Moe, V. F. 89–91, 90–1, 93, 94
Molaison, H. 40
monkey and human, mirror neurons in 81, 83–7, 130, 138, 140–51
mood swings 26
Moore, Jonaven 119
Morgan, William J. 104–5
Morse, S. 24, 29, 31
motor actions 82, 83, 85, 88–94, 93, 139, 140
motor image 83
motor knowledge 83, 91
motor system 82
motor theory of mind 89–90
Munenori, Yagyū 122
mushin and flow 119–20
Myin, E. 114, 116–18, 121, 124

National Football League (NFL) 3, 11, 13, 23–8, 31, 32, 39
Navratilova, M. 112, 116
neural Darwinism 83
neural transmission rate 129–31
neuro-centrism 73
neurocorrelates 30
Neuro - defence research programme 72
neurodoping impact 57–8
neuroethical analyses 105
neuroethical approach 104
neuroexuberance 24, 29
neurological condition 43
neuromania 24, 29
neuropathologies of the self 106
neuroregulatory mechanisms 31
neuroscience 1–2, 54; discourses 25; knowledge 3, 24–33
neuroscientific research 84
neurotechniques 56
new public health 30
NFL *see* National Football League (NFL)
Noe, A. 114
Nomenclature for performance achievement levels (NPAL) 50
noradrenalin (norepinephrine) 55
normalization of concussion 27
Nutt, Amy Ellis 42

observational learning 84–5
Olympic motto "Citius, Altius, Fortius" 59
Omalu, Bennet 25, 26
one fruitful approach 101, 102
On Liberty 8
On the old saw: that may be right in theory but it won't work in practice (Kant) 12
Open: An Autobiography (Agassi) 41
O'Regan, J. 114
Ostlund, Ruben 68

INDEX

The Paradoxical Brain (Kapur) 43
paradoxical cognitive phenomena 43
parasympathetic Autonomic Nervous System 73
Parfit, D. 13
parietal lobe 82
Park, Fenway 134
Parry, J. 8
paternalistic policies 9
Paul, L. A. 43–5, 46n21
Pavlov, Ivan 73
personality changes 26
personal knowledge 91
Petersen, A. 30
physical and psychological enhancements 50–2
physical injury 38
physicalism 38
Pickersgill, M. 28, 29
pitch caller 133–5
pitchers 108, 128–32, 134
pitch framing 131
PITCHf/x tracking system 128–31, 132, 134, 135
pitching dominance 133–4
Pitching Triple Crowns 128
pitch-tracking technology 134
plasticity 39
Platos academy 66
Polanyi, M. 91, 93, 94
politico-philosophy 27
Pontius, Annelise A. 2
Popper, Karl 26
posterior motor areas 82
postmortem examination 26, 38–9
post-traumatic growth 43
posttraumatic stress disorder (PTSD) 40, 42
potential objections and responses 42–4
power and persuasion 16
precariousness 14
preemptive paternalism 11, 13
'pre-emptive paternalism' 10
prima facie 42
procedural paternalism 18
processing speed reduction 26
program of embodied cognition 82
prohibition exists 16–18
property dualism 38
psychological shock 38
punch drunk syndrome 25

Racine, E. 2, 33n1
radical enactive cognition (REC) 114, 115, 124
Rasmussen's syndrome 39
rationality 13, 14, 27, 28
reaching and grasping 82
relational embeddedness 107
reliability 30
Re: NFL Players 28

resilience 42, 50
Rickers, K. 102
risk and anticipation 28–9
risk-taking practices 27
Rizzolatti, G. 81–94, 95n3, 138, 140–4
Robinson, L. 41
Roelofs, Karin 72
Roentgen, William Conrad 29
Rose, N. 24, 25, 27–30
Rosenberg, D. 16
Roskies, Adina 2
Rowlands, M. 117
Rusie, A. 133
Ryan, J. 41
Ryle, G. 100

saccades 130
Safai, P. 27
Safire, William 2
Sailors, Pam 8–11; Mills slavery analogy 11–18
Sampedro, Alberto Carrio 3
Sandel, M. 108
Sartre, Jean-Paul 3, 66–8, 72, 75–7
Satel, S. 30
Sawyer, K. 107
scaffold, behavioural 72–4
Scenes from a Marriage (Bergman) 75
Schonsheck, J. 12
scientific evidence 59–60
scientific research 66–7
scientific uncertainty 26–8
Searle, J. 91, 94
sed contra est 66
Seeing Stars: Emotional Trauma in Athlete Retirement: Contexts, Intersections, and Explorations (Tinley) 41
self-exposure 77
Sey, J. 41
shooting 119, 131
Siegfried 73
Simon, Robert L. 13
single-cell recordings 81–3, 86, 88, 95n1
Sinigaglia, C. 82–5, 87, 89–93, 140
skiing 3, 70, 74; and philosophy 65–9, 75–7
ski resort 65–6, 68, 71, 72, 75, 76
slave analogy 14
slavery and playing football/boxing 13
sleeplessness 26
sliding 67, 68
social brain 101
social synapse 37
socio-technical ambiance 72–4
soft and hard paternalism (Dworkin) 8
soft paternalism *vs.* hard paternalism 9
'soft' paternalistic interventions 9
spiritual approach 107

sport: cognitive abilities in 52–4; concussions *see* concussion; criticisms of 57–9; emotional trauma and 41–2; medicine 17; performance 53; philosophy 75–7, 90–3; practices 27; psychology 52; risk and the biopolitics 26–8
sporting skills 92
sport related concussion (SRC) 16–19
stathmin gene (STMN1) 73
Strasser, M. 15
stress and abuse 42
stress functions 42
strictly congruent mirror neuron 83
subsequent non-mechanical effect 17
substance dualism 38
Sullivan, P. 102
superior temporal sulcus (STS) 86
sympathetic Autonomic Nervous System 73
synaptic connections 24–5

Takuan Sōhō 122
Tamorri, S. 54
tau protein 26, 31, 38–9
TCS *see* transcranial current brain stimulation (TCS)
tDCS *see* transcranial direct current brain stimulation (tDCS)
team chemistry: definition 100–1; working 101–2
team flow 106–7
team spirit: definition 99–101; ethics 102–3; flow 106–7; liberal individualism and limits 103–6; performance 108–9; and team chemistry 99, 106, 107; working 101–2
technicity of skiing 68
technology 51
The Teenage Brain: A Neuroscientist's Guide to Raising Adolescents and Young Adults 42
teleosemantics 117
temporal neutrality 14
tetrahydropregnanolone (THP) hormone 42
theories of mind 112–13
theories of sport 57–9
theory–theory of mind 89–90
Thomaschke, R. 84
Thompson, E. 82, 114
Tinley, S. 41

tragic form of art 68
transcranial current brain stimulation (TCS) 56–7
transcranial direct current brain stimulation (tDCS) 56–7
transcranial magnetic stimulation (TMS) 56
traumatic brain injury (TBI) 45n2
traumatic interrelationship 42
trauma victims 39
triangulation 69, 75–7
Trivino, Jose Luis Perez 3
Turist 3, 68, 70–3, 75–7
Turner, Keith 24

umpires 4, 127–35
uncertainty 18, 26–8
unprecedented metamorphosis 43–4
unsafe bridge (Mill) 9

van der Kolk, Bessel 40
Van der Veer, D. 9
victims of trauma 39
Vincent, N. 45n7
violenti non fit injuria 9
visual and motor responses 83
visual system 83, 130
Vogt, S. 84
voluntary dominance 11
voluntary slavery 10
vulnerability 2, 6, 14, 17, 41, 76
vulnerable brain 17

Webster, Mike 25
Weiss, C. 26
Williams, Ted 129
wingsuit flying 15
Wooden, John 102
Wooden, Shawn 24
World Anti-Doping Agency 50, 58
World Health Organization (WHO) 50
wushin 119

X-rays 25, 29

zanshin 119
Zola, Emile 70–1
Zwart, Hub 3

For Product Safety Concerns and Information please contact our EU
representative GPSR@taylorandfrancis.com
Taylor & Francis Verlag GmbH, Kaufingerstraße 24, 80331 München, Germany

www.ingramcontent.com/pod-product-compliance
Ingram Content Group UK Ltd.
Pitfield, Milton Keynes, MK11 3LW, UK
UKHW031043080625
459435UK00013B/545